# Advances in *Candida Albicans*

*Edited by Xinhui Wang*

Published in London, United Kingdom

IntechOpen

*Supporting open minds since 2005*

Advances in *Candida Albicans*
http://dx.doi.org/10.5772/intechopen.87410
Edited by Xinhui Wang

Part of IntechOpen Book Series: Infectious Diseases, Volume 9
Book Series Editor: Shailendra K. Saxena

Contributors
Humam Kasem Kasem Hussein, Estela Ruiz Baca, Ana Lilia Martinez-Rocha, Rosa Isela Arredondo Sánchez, Karina Corral-Pérez, Angélica López-Rodríguez, Iván Meneses-Morales, Víctor Manuel Ayala-García, Snigdha Pattnaik, Laxmidhar Maharana, Manoj Kumar Sethi, Sandra Widaty, Eliza Miranda, Caroline Oktarina, Ahmad Ibrahim, Rashi Verma, Dibyabhaba Pradhan, Harpreet Singh, Arun Kumar Jain, Luqman Ahmad Khan



First published in London, United Kingdom, 2021 by IntechOpen
IntechOpen is the global imprint of INTECHOPEN LIMITED, registered in England and Wales, registration number: 11086078, 5 Princes Gate Court, London, SW7 2QJ, United Kingdom
Printed in Croatia

British Library Cataloguing-in-Publication Data
A catalogue record for this book is available from the British Library

Additional hard and PDF copies can be obtained from orders@intechopen.com

Advances in *Candida Albicans*
Edited by Xinhui Wang
p. cm.
Print ISBN 978-1-83969-181-2
Online ISBN 978-1-83969-182-9
eBook (PDF) ISBN 978-1-83969-183-6
ISSN 2631-6188

For EU product safety concerns:
IN TECH d.o.o., Prolaz Marije Krucifikse Kozulić 3, 51000 Rijeka, Croatia, info@intechopen.com or visit our website at intechopen.com.

IntechOpen Book Series

# Infectious Diseases

Volume 9

Dr. Xinhui Wang obtained a Ph.D. from the University of Twente, Netherlands, in 2008. Over the years, he has worked in bioinformatics at the Norwegian University of Science and Technology (NTNU), Norway, University Medical Center Utrecht Utrecht (UMC), Netherlands, Radboud University Medical Center, Netherlands, and Academic Medical Center (AMC), Netherlands. His research focuses on multi-clinical omics data analysis, including metagenomics, genomics, immunomics, and transcriptomics data analysis. He has helped develop many statistical tools and machine learning approaches for all kinds of clinical data analysis. He has served as a committee member of more than ten conferences and a reviewer of ten academic peer-reviewed journals.

**Editor of Volume 9:**
**Xinhui Wang**
School of Computer Science, Qinghai Normal University,
Xining, China

**Book Series Editor:**
**Shailendra K. Saxena**
King George's Medical University

# Scope of the Series

The series will give a most comprehensive overview of recent trends in various infectious diseases (as per the most recent Baltimore classification), as well as general concepts of infections, immunopathology, diagnosis, treatment, epidemiology and etiology to current clinical recommendations in management of infectious diseases, highlighting the ongoing issues, recent advances, with future directions in diagnostic approaches and therapeutic strategies. This book series will focus on various aspects and properties of infectious diseases whose deep understanding is very important for safeguarding human race from more loss of resources and economies due to pathogens.

# Contents

# Preface

*Candida albicans*, a fungal pathobiont, is the major component of the microbiota communities in healthy adults. It resides in the host's gastrointestinal tract and mouth. *C. albicans* can become pathogenic via overgrowth of the fungus under a variety of conditions. Infection caused by *C. albicans* can form a biofilm that is resistant to antifungal therapeutics and the host immune system. The epithelial cells in mucosa help develop elaborate immune responses again *C. albicans* infection. Genetic mutations play an important role in the virulence of *C. albicans*. Whole-genome sequencing has revealed more identifications of population structure, epidemiological investigations, and phylogenetic analyses of *Candida* species.

This book reviews recent knowledge and the latest research on *C. albicans*, including the mechanism of candidiasis infection, host response, antifungal strategies, biofilms, genetics, and molecular epidemiology of immune responses.

Chapter 1 examines several factors of *C. albicans* pathogenesis. It surveys all the underlying variables and components of pathogenesis to improve understanding of these factors' effects on modulate virulence and consequent infection.

Chapter 2 discusses the molecular mechanisms of resistance to antifungal agents at the molecular level described in *C. albicans*. The information presented may be helpful for discovering new antifungal agents or targets to combat candidiasis.

Chapter 3 reviews current advances in model construction, target identification, and validation. It presents several examples of successful metabolic model construction and these models' utility in rational drug design.

Chapter 4 examines the effect of cell-mediated (T cells) and immune cells (macrophages, neutrophils, and natural killer cells) on *C. albicans* infection. The chapter adds to the understanding of immune responses and antibody-mediated responses fighting infection.

Chapter 5 discusses onychomycosis, a common fungal infection affecting nails. Caused by *C. albicans*, onychomycosis is frequently associated with local or systemic immune disturbances. Microscopic examination and fungal cultures are the gold standard methods for diagnosing onychomycosis.

Chapter 6 discusses *C. albicans* and the risk of miscarriage. Excessive growth of *C. albicans* can cause vulvovaginal candidiasis, which, if chronic and recurrent in pregnant persons, may contribute to spontaneous abortion or miscarriage.

Understanding the mechanism of *C. albicans* infection can aid in developing proper treatment and discovering novel drugs.

**Xinhui Wang**
School of Computer Science,
Qinghai Normal University,
Xining, China

Section 1

# Pathogenesis Factors of *Candida albicans*

Chapter 1

# Pathogenicity Mechanism of *Candida albicans*

*Snigdha Pattnaik, Laxmidhar Maharana and Manoj Sethi*


## Abstract

In normal human microbiome, the polymorphic fungus *Candida albicans* is a crucial member. *C. albicans* resides mostly in individual as harmless commensal life. In specific situations, however, *C. albicans* can cause diseases that cause contaminations of the skin to life-threatening fundamental contaminations. Pathogenesis of Candida species is contributed by multiple factors. Some of the major contributors are enlisted here. These include host pathogen interaction, receptors molecule like TLR recognition, TLR signaling, C type lectin receptors, Dectin 1,2 and 3, mannose receptor, mincle, DC sign, Nod-Like Receptors (NLRs) and inflammasomes, soluble molecules in candida recognition, cellular responses to candida such as neutrophils, macrophages. This chapter enlightens all the components of candida pathogenicity by the assessment of Candida species pathogenic determinants. All together these will explain the current knowledge about how these determinant factors and receptors modulate virulence as well as consequent infection. Better understanding of candida pathogenicity mechanism can be the resultant of better treatment guidelines along with development of novel antifungal agents. Overall, in this review we present an update in the current understanding of the insight of pathogenicity mechanisms in this important human pathogen.


**Keywords:** Pathogenicity, *C. albicans*, TLR, receptor, lectin

## 1. Introduction

Candida is a diploid parasite that as often as possible causes mucosal and fundamental contaminations in people [1]. Candida species can colonize a few particular anatomical locales. Greater part of diseases by commensal microorganisms comes from endogenous colonization. Notwithstanding, exogenous pollution, for example, diseases communicated through emergency clinic workers, medical clinic air, and biofilm-debased intrusive gadgets like catheters, can likewise happen [2–4]. Diseases brought about by Candida can be delegated shallow, cutaneous, mucosal, and fundamental infection. At the point when Candida spp. taint the oral cavity, skin, genitalia, respiratory framework, and the remainder of the gastrointestinal lot, the disease is delegated the shallow sort. Intrusive candidiasis is a disease portrayed with very extreme conditions, for example, candidemia, meningitis (influencing the mind), and endocarditis (influencing the heart) [5]. In hospitalized patients and those with bedraggled safe framework, intrusive contamination is a huge reason for dismalness and mortality along with increased frequency as well as pervasiveness rates.

Candida species pathogenesis is a complex cycle including numerous instruments and pathways. It is likewise a mind boggling and multifactorial system, including highlights of both the host and the microorganism [6]. For contamination to be set up, the pioneering microorganism should avoid, duplicate in the host climate, and make do in the safe arrangement of the host. The living being must likewise have the option to scatter to other body tissues and organs, most particularly in foundational disease [7]. Problem in skin or gastrointestinal boundaries can prompt dispersed or profound organ candidiasis. In more significant circumstances, circulatory system intrusion may some time possible which hence will disperse to various organs of the body.

Candida contaminations in a great many people are asymptomatic. This is because of the capacity of the immunological framework to checkmate the life form as it endeavors to spread in the body. In any case, consumption in resistant systemor changes in microbiota balance, combined with different elements, can work with the spread of Candida which is regularly deadly in 42% of announced cases [8–10]. *C. albicans* is answerable for about half of candidiasis and non-albicans Candida species are liable for the rest of the Candida contamination. Disease brought about by several other species of candida are of extraordinary concern. A portion of these non-albicans Candida species are presently viewed as arising artful microbes [11]. Forestalling Candida contaminations for the most part brought about by Candida species is a developing test in human medication. Indeed, even with the accessibility and utilization of antifungal prescription, scattered candidiasis is went with high death rate (around 40–60%), helpless conclusion, and unseemly illness the board. The overall clinical show of the patient likewise adds to the expansion in death rate. Protection from antifungal medications is not, at this point another issue. Indeed, even among people that have not been presented to anti-infection agents, obstruction has been accounted for [12]. Candida is one of the main sources of mucosal contaminations in sound people for now days. It additionally causes initial diseases particularly in immunosuppressive patients, regardless of its status as a commensal microorganism [13]. Truth be told, candidiasis is viewed as the third to fourth most regular infection in medical care offices inside the USA and even all around the world [14].

As anyone might expect, it is the destructiveness and pathogenic qualities and components that have gotten the most consideration from specialists throughout the long term. As of late, much have been found out about the components of Candida pathogenesis. Studies have shown that at the core of the capacity of Candida to multiply, change from non-destructiveness commensal to pioneering pathogenic organism and build up disease in the host lie profoundly interconnected elements made out of transcriptional circuits, morphology-related/harmfulness encoding qualities, metabolic versatility, genome pliancy, phenotypic exchanging, biofilm arrangement, tissue harming extracellular hydrolytic catalysts, and a few different variables that work with destructiveness and pathogenesis in Candida species [15]. Changes in ecological pH, vigorous supplement procurement framework, escape from phagocytosis, avoidance from have insusceptible framework, have microbiome coaggregation, protection from antifungal specialists, and the capacity to productively react to numerous anxieties are other crucial characteristics that upgrade endurance and pathogenesis.

In order to be capable of inducing such a diversity of infections *C. albicans* can live in several anatomically discrete sites and translates several virulence factors. The phenomenon of phenotypic converting from yeast- to filament-growth is just one, but critical, factor that contributes to the virulence of *C. albicans*. It offers a basis for activating different receptors leading to diverse immune responses. Other virulence factors of *C. albicans* contain adhesion factors, thigmotropism and secretion of several hydrolytic enzymes, such as lipase, phospholipase, and proteinase. During the past few years it has become increasingly clear that PRRs are vital for the host response to *C. albicans*, with various TLRs and LRs having distinctive roles in innate immunity. Each ligand–receptor system activates specific intracellular signaling pathways, which in turn leads to modulation of various components of the host immune response. While a few receptors, like TLR4, dectin 1 and the MR, apply an all the more favorable to fiery job, others employ immunosuppressive impacts (for instance, TLR2, CR3 and FcγR). After disclosure and characterized clarification of the part of TLRs in parasitic acknowledgment, further investigations have explained the job of the C-type lectin receptors with an emphasis basically on dectin-1 and dectin-2. The presence of various relationships among all of the components that guide the establishment of pollutions is an undeniable component in the pathogenesis of Candida species. This chapter is precisely based on the mechanisms of Candida pathogenesis with emphasis on the virulence factors mostly the important receptors and pathogenic determinants.

## 2. Pathogenicity mechanism of Candida species

### 2.1 Infection

The pathogenicity of *C. albicans* is identified with its change between the commensal yeast structure and the obtrusive hyphal shape [16]. Upon have cell connection, thigmotropism (contact detecting) triggers *C. albicans* filamentation. This allows the creature to infiltrate further into the host tissues through extracellular compound emission [17]. The capacity of Candida to change over from yeast to hyphae stage or hyphae to yeast stage is named dimorphism. Every one of these periods of development is crucial for harmfulness and pathogenicity as it impacts how Candida gets away from the resistant framework. Yeast and fiber (hyphae) structures assume autonomous parts during scattered candidiasis. While the yeast structures engaged with scattering, the hyphal (filamentous) structure is associated with tissue intrusion and pathogenesis [18]. Candida species should have the option to adequately colonize its host and moreover adjust to assortments of unessential requirements like temperature, oxygen, pH, carbon dioxide, and diverse negative organic conditions, for example, carbon source, supplement accessibility, the immunological framework, and other existing together bacterial and contagious cells inside the specialty [19, 20]. Positive reaction to those imperatives has a quick impact in transformation and advancement of Candida harmfulness and pathogenicity. Before receptor-intervened epithelial acknowledgment by Candida species, a few flagging pathways are actuated. Temperature change, supplement starvation, oxidative pressure, osmotic pressure, and pH detecting trigger mitogen-enacted protein kinase, pathways based on CAMP, transduction of Rim-101, along with surprisingly hereditary mechanisms that constantly instigate numerous qualities. Most of the induced characteristics are connected with filamentous turn of events and biofilm plan. While assorted hereditary pathways transduce shifts from yeast to hyphae or hyphae to yeast stage, distinctive ecological signals emphatically and contrarily regulate morphology-related cell surface exchanging [21]. The flagging and variation pathways assume pivotal parts in different physiological and cell measures engaged with the Candida species pathogenicity as demonstrated in **Table 1**.

The greater part of the flagging pathways are amazingly fundamental for protecting Candida spp. against immunological assault [40]. They assume different parts in the declaration of morphology related qualities. The co-articulation of morphology-connected proteins brings about synergistic association among

| S/no. | Pathways | Functions | Reference(s) |
|---|---|---|---|
| 1 | Mitogen-activated protein kinase (MAPK) pathways | Important regulator of morphogenesis. | [22, 23] |
| | | Involved in sensing and transmitting stress signals and other environmental signals | |
| | | Three main MAP kinase pathways are the following:<br>a. Mkc1- controls cellular integrity, invasive growth, cell wall biogenesis, and formation of biofilm<br>b. Hog1- mediates response to thermal, osmotic, and oxidative stress. Controls cell wall formationand morphogenesis. Under osmotic stress, its activation leads to glycerol accumulation.<br>c. Cek1- it mediates mating and hyphae formation and is also involved in adaptation to boththermal and nutrient stresses. | [24]. |
| 2 | Ras-CAMP-PKA pathways | Regulate adhesion, dimorphism. Also involved in the formation of biofilms. | [23, 25, 26] |
| | | Control hyphal formation and white-to-opaque change | [27, 28] |
| | | Involved in drug tolerance and in the maintenance of cell wall integrity | |
| 3 | RIM 101 signal transduction | Enables *Candida albicans* to sense pH changes, thus mediate pH-dependent responses | [29] |
| 4 | Stress response pathways | Contribute to virulence and pathogenesis<br>Facilitate adaptation to ever-changing environmental conditions.<br>Protect against host-derived stresses | [30] |
| 5 | Ergosterol biosynthetic pathways | Link between hyphae formation and virulence in *C. albicans* | [31] |
| | | Enhance cell adhesion and damage to the tissues | [32] |
| | | ERG3 and ERG11 play major roles in azole drug resistance; thus, it is the target of fluconazoleantifungals | |
| 6 | Genome plasticity | Triggers adaptation to fluctuating host environment.<br>Leads to the generation of recombinant progeny with increased fitness.<br>Induces natural mutations that alter the balance between commensalism and pathogenicity. | [33, 34] |
| | | Facilitates resistance to stressors including antifungal agents and pathogenicity during systemic and mucosal infections | [35] |
| | | Triggers polarized filamentous growth<br>Involved in the generation/evolution of new pathotypes or strains<br>Enhances the utilization of several nutrients.<br>Facilitates Candida growth rate, as well as its morphology and behaviors at the host interface | |
| 7 | Calcium-calcineurin pathways | Major mediator of stress responses | [36] |
| | | Essential for survival in the presence of stressors | [37] |
| | | Play crucial roles in virulence | [38, 39] |

**Table 1.**
*Major pathogenicity inducing pathways/responses in Candida species.*

quality items fundamental for biofilm foundation and development inside the host [41]. Along these lines, for hindering Candida endurance in have tissues, impedance with Candida species capacity to incorporate quality articulation to changes in morphology could be surely a potential restorative technique [42]. Also, distinguishing flagging segments saved among Candida species is vital for recognizing potential medication targets. During the interaction of pathogenesis, actuated endocytosis happens. It for the most part happens inside 4 h of starting contact to epithelial cell. Candida uses prompted endocytosis to sidestep invulnerable acknowledgment. The acknowledgment of invasins communicated on the contagious cell surface triggers prompted endocytosis. Until this point, only A1s3p and Ssa1p (invasins) are known for *C. albicans*. In a murine model of oropharyngeal candidiasis revealed by Sun et al., Als3 and Ssa1 freaks displayed diminished grip and intrusion of cells of epithelium [43]. Free of the cellular receptor of epithelium, instigated endocytosis can likewise happen. This is conceivable through the association of the host epithelial cell epidermal development factor receptor with the invasins of candida cell. Post actuated endocytosis, discharged harmfulness factors by pathogens to improve capacity to enter to surface of mucosa. The oral and vaginal mucosa, which are terminally separated and non-proliferative, are made out of delineated layers more averse to work with intrusion of parasites by means of initiated endocytosis. Candida species should use an elective course to attack a tissue less inclined to help disguise in a cycle called dynamic infiltration. Dynamic infiltration interceded through hyphae augmentation (constrained by Ume6 and Eed1) is a contagious actuated cycle that needs reasonable parasitic hyphae [44]. Actual powers, attachment, and hydrolytic chemicals like SAP additionally assume a part. *C. albicans* uses dynamic entrance as the underlying way to attack the furthest layers of the epithelium in vivo. Be that as it may, prompted endocytosis could likewise be obvious of additional upgraded attack once the fundamental proliferative layers of the epithelium have been gotten to by the growth. Along these lines, both dynamic infiltration and initiated endocytosis are unthinkingly noticeable systems required for disease foundation through mucosal boundaries in vivo. When all is said in done, the pathogenesis of Candida begins with colonization, shallow disease, and profound situated contamination before spread contamination. The overall strides in tissue intrusion by *C. albicans* incorporate in the following stages.

a. Adhesion to the cellular epithelium.

b. Colonization.

c. Penetration to epithelium/hyphal invasion.

d. Dissemination of vasculature.

e. Endothelial colonization/penetration.

Systemic candida infection only occurs by immune system escape than vasculature penetration and invading the blood components. Entry to the bloodstream occurs via two routes:

a. Natural routes.

b. Artificial routes.

Above subsequent course is worked with biofilm arrangement as pathogens can get away and invade the blood. For Candida to endure and spread in the blood, various qualities are upregulated: qualities engaged with protein amalgamation, glycolytic cycle, glycolysis, and reaction to oxidative pressure. The presence of Candida in the blood prompts a condition called candidemia. From the blood, the yeast is dispersed to different fundamental organs in the body where it causes foundational contaminations. Dispersed candidiasis is profoundly worked with by extracellular hydrolytic compounds, adhesins, phenotypic exchanging, and cytolytic proteins. Candida in the blood can likewise bring about candiduria by antegrade contamination. Albeit most diseases include biofilm arrangement, a few contaminations can happen without the development of biofilm. Indeed, hyphae development and development are the beginning stages in the pathogenicity of Candida species, with the exception of *C. glabrata* that does not shape hyphae. It is notable that few qualities straightforwardly or by implication incited by natural irritations trigger hyphae arrangement.

Notwithstanding, questions actually remain with respect to the instruments controlling its union, the receptors, and its carrier. In outline, the exchanging of Candida spp. from commensal to artful microbe is ascribed to destructiveness factors that are specifically communicated under reasonable inclining conditions. The majority of these destructiveness factors are under close guideline. More examinations in their administrative instruments could be fundamental in the mission for new antifungal specialists. **Figure 1** is the significant organization of Candida destructiveness and pathogenesis showing the associations between the different pathogenic determinants and harmful variables.

### 2.2 Host response to Candida species

Host insusceptible acknowledgment of Candida happens through a few instruments involving intrinsic and versatile insusceptibility. The versatile insusceptible framework perceives explicit antigenic moieties, prompting the advancement of a focused on safe reaction. Interestingly, inborn insusceptible acknowledgment is vague and wide and is the primary line of host protection against possibly hazardous organisms. These vague reactions are promptly endless supply of an organism in a pre-modified design and assume a fundamental part in controlling contagious weights and forestalling infection. Natural invulnerability includes a progression of dissolvable (supplement) and cell (neutrophil, macrophage) parts that act in show to keep by far most of microbes from setting up an intrusive disease. Further, it has become progressively clear that these reactions capacity to enact versatile insusceptibility just as acting along with other homeostatic cycles to give further security. Natural invulnerable acknowledgment of Candida happens through the acknowledgment of microorganism related atomic examples (PAMPs).

PAMPs are themes or particles that are regular between various sorts of growths. In contrast to antigens, individual PAMPs are not explicit to a solitary Candida animal variety but instead are divided among various species and contagious genera. These microbial PAMPs are perceived by have germline encoded design acknowledgment receptors (PRRs) [45] and give a pre-customized method of parasitic acknowledgment, taking into consideration moment acknowledgment of normal contagious parts. Most of contagious PAMPs are cell divider related and incorporate $\beta$-glucans, *N*-and *O*-connected mannans, and phospholipomannans [46]. These are perceived by three key PRR families: cost like receptors (TLRs), C-type lectin receptors (CLRs), and nucleotide-restricting area leucine-rich receptors (NLRs) [46–52]. Dendritic cells, monocytes, macrophages, polymorphonuclear leukocytes (PMNs), Tcells, Bcells, and epithelial cells all transmit PRRs on a surface level, in

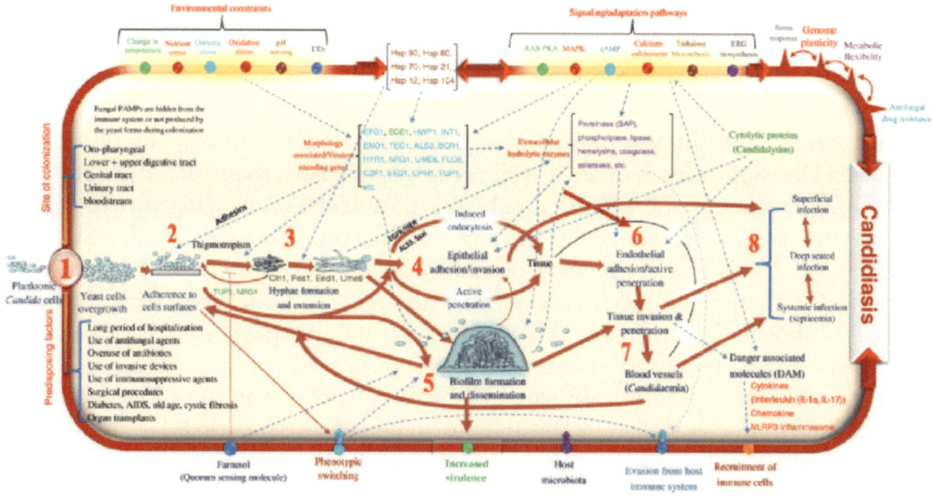


**Figure 1.**
*Simplified diagram illustrating the network of Candida virulence and pathogenicity. (1) planktonic yeast cells attach to surfaces. Favorable conditions facilitate overgrowth; adherence (2): The cells attach to host cells via adhesins; hyphae formation/extension (3): Environmental constrains induce the HSPs, signaling and adaptation pathways which induce morphology-associated genes. The formation of the hyphae marks the beginning of Candida pathogenesis. Epithelial/endothelial adhesion/invasion (4 and 6): This is facilitated by hydrolytic enzymes and it is achieved via two ways: Induced endocytosis and active penetration. Some species such as C. glabrata do not form hyphae; rather, they form biofilms (5) prior to the establishment of infection. Destruction of epithelial and mucosal surfaces by the enzymes and cytolytic proteins gives rise to different types of candidiasis (8). Yeast cells can enter the blood (7) and then disseminate to the vital organs where they establish new biofilms. Infections associated with biofilms are of great clinical significance. Major Candida infections include vulvovaginal, oropharyngeal, and gastrointestinal candidiasis, candidemia, candiduria, and intra-abdominal candidiasis. Key: Dashed lines: Signals and inductions; single-headed thick dark red arrow: Major route of Candida pathogenesis; curved double-arrow connector: Interaction/association between factors; T-shaped thin red line: Inhibitory signal. The pool of virulence encoding genes house both the genes involved in hyphae and biofilm formation and other vital processes crucial for pathogenesis.*

endosomes or in the cytoplasm of host cells. Sanctioning of these PRRs by PAMPs prompts setting off of intracellular hailing pathways, as MAPK (mitogen-started protein kinase) and NF-κB (nuclear factor kappa-light-chain-enhancer of incited B cells) pathways, and finally to further developed record of countless characteristics drew in with have safe protections, including chemokines, cytokines, provocative center individuals, and antimicrobial peptides. Appropriately, PRRs are fundamental center individuals among intrinsic and adaptable safe responses.

## 3. Receptor molecules in Candida recognition

While comparing the human genome with murine genome; human genome encodes for ten TLR characteristics (TLR1–10) and murine genome encodes 12 i.e. TLR1 to TLR9 and TLR11 to TLR13. Each TLRs depicted as transmembrane type-1receptors having an enriched lucine extracellularly intermittent region which sees target PAMP and a Toll/interleukin-1 receptor-(TIR-) space containing cytoplasmic region that imparts the institution stimuli, which having closeness to the sort 1 interleukin-1 (IL-1) receptor. TLR family is a developmentally monitored gathering of PRRs that react to an assortment of bacterial, viral, and contagious PAMPs just as some endogenous components delivered when have cells are harmed. The extracellular areas of TLRs perceive an assortment of microbial PAMPs, including lipopolysaccharide (LPS), peptidoglycan, proteins (counting triacylated proteins and flagellin), and changed nucleic acids [53–58].

## 3.1 Toll like receptors

### *3.1.1 TLR recognition of Candida*

Key part for TLRs in host protection against fungal infection was initially identified when Drosophila inadequate in Toll receptor were seen to profoundly helpless to A. fumigatus disease [59]. Therefore by far most of the underlying antifungal insusceptibility research focused on how contagious cells were perceived. This provoked the distinctive verification of a couple of PRRs related with affirmation of different cell divider polysaccharides of parasites and *C. albicans* explicitly, including TLR2 (phospholipomannan), TLR4 (*O*-associated mannan), and mannose receptor (MR) (*N*-associated mannan) [46, 48, 60].

At last, these investigations finished in the disclosure of another PRR, dectin-1 (dendritic cell associated C-type lectin-1), who perceives parasitic $\beta$-1,3 glucan [61]. Outstandingly, these parasitic PRRs can work both freely and related to each other. For instance, dectin-1 and TLR2 act additionally to perceive contagious yeasts, with dectin-1 prompting phagocytosis while TLR2 initiates cytokine creation [62–64]. Dectin-1 likewise synergises with TLR4 flagging [64]. Moreover, TLR1 and TLR6 structure heterodimers with TLR2 [65] however do not seem to assume a significant part in *C. albicans* acknowledgment in a mouse model of intrusive candidiasis [66]. Obviously depending upon the coreceptor included, coligation of TLR2 may either update TLR2-subordinate responses [67] or change its PAMPs distinction concerning the circumstance with galectin-3 [68].

Even so these are standard receptors utilized by macrophages and neutrophils to see *C. albicans*, various receptors have moreover been perceived inclusive of dectin-2 [69], mincle (macrophage inducible CTL) [70], Dendritic cell specific intercellular grasp particle 3- getting nonintegrin (DC-SIGN ) [71, 72], and galectin-3 [68]. The piece of these PRRs is correct now not totally settled; regardless, dectin-2 and DC-SIGN are perceived to assume a significant part in the acknowledgment of high mannose structures [73] and galectin-3 in the acknowledgment of $\beta$-1,2 mannosides [68].

Curiously, galectin-3 coimmunoprecipitates accompanied by dectin-1 [74], which recommends that galectin-3 can work with associations among TLR2 and dectin-1 flagging. TLR acknowledgment of other medicinally significant growths have likewise been concentrated yet are less very much described, despite the fact that apparently TLR3 perceives A. fumigatus conidia and TLR4 perceives *Cryptococcus neoformans* glucuronoxylomannan, with TLR9 perceiving *A. fumigatus, C. albicans* and *C. neoformans* [75].

### *3.1.2 TLR signaling*

PAMP acknowledgment of TLRs brings about enactment of flagging cascade intracellularly (**Figure 2**) through connection of the cytoplasmic TIR spaces with various connector proteins: myeloid separation essential reaction quality (88) (MyD88), MyD88-connector like (MAL), TIR-area containing connector initiating interferon-$\beta$ (TRIF), and TRIF-related connector atom (TRAM) [53–58, 76–79]. This TLR-adapter interaction ends up in the activation of the IRAK ( IL-I receptor associated kinase) proteins and TRAF6 (TNF receptor associated factor-6). As a result it ends up in activation of the main signaling pathways together with NF-$\kappa$B, MAPK, and IRF (interferon regulative factor) pathways. MAPK activation contains 3 alleyways: p38, JNK (c-Jun N-terminal kinase), and ERKI/2 (extracellular signal-regulated kinaseI/2). Finally, signaling pathway induction ends up in the activation

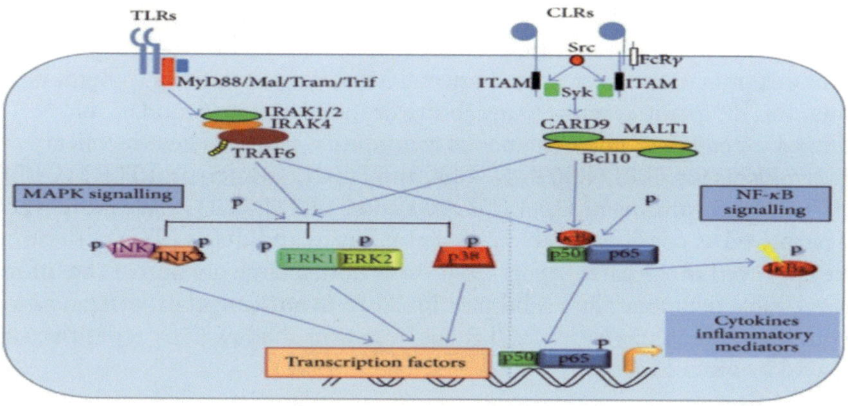


**Figure 2.**
*Signal pathway activation by TLRs and CLRs. TLRs and CLRs activate MAPK and NF-κB signal pathways to varying extents, thereby allowing different innate immune responses to be generated. TLRs utilize TIR-domain containing adapter proteins such as MyD88, mal, TRAM, and TRIF. CLRs signal using ITAM domains within their cytoplasmic region (e.g., dectin-1) or associate with an ITAM-containing transducing protein (e.g., dectin-2 with FcRγ). Dectin-1 utilizes Src kinases and Syk kinase to activate a complex containing CARD9, MALT1, and Bcl10 to activate the downstream signal pathways. Figure adapted from [47].*

and nuclear localisation of transcription factors as well as NF-κB, AP-I (activating macromolecule I), and IRF-3 and IRF-7. the result of this activation cascade is to induce organic phenomenon and secretion of varied proteins concerned in immune defense as well as cytokines, chemokines, antimicrobial peptides, and alternative inflammatory mediators, all of that operate to stimulate innate and reconciling responses of immune system. It thought to be noted that the overwhelming majority of studies shaping the TLR-mediated pathways are performed victimization myeloid or humor cells, however elaborated analysis of TLR- mediated pathways in alternative cell varieties, and specifically animal tissue cells, could nonetheless establish novel and strange mechanisms of infectious agent (fungal) recognition and management at membrane surfaces.

### *3.1.3 Role of TLRs during Candida infection*

Although animals missing the TLR signaling adaptor protein MyD88 are vulnerable to fungal infection [46, 80–82], the exact role of particular TLR receptors in fighting Candida infections is unclear. This is most likely because of contrasts in examination plan, where diverse contagious species, morphotypes, and courses of contamination have been surveyed [52]. Thusly, contemplates utilizing TLR knockout mice have uncovered critical contrasts in the putative jobs of various TLRs in fundamental or mucosal insusceptible reactions against contagious contaminations [83]. For instance, while a few examinations demonstrate that TLR2 and TLR4 impact vulnerability to murine scattered candidiasis [82, 84–86], not all investigations support this attestation [87, 88]. TLR7 might be needed for parasitic RNA acknowledgment in the autophagosome, which is needed for IFN-β discharge and is related with delayed *C. glabrata* contamination [89]. TLR9 perceives *C. albicans* DNA (unmethylated CpG arrangements) bringing about cytokine creation in dendritic cells [90]; notwithstanding, TLR9 knockout mice do not seem, by all accounts, to be more helpless to *C. albicans* contamination, notwithstanding delivering diminished degrees of IL-I2 and expanded measures of IL-4 and IL-I0 [82, 90–92]. Outstandingly, explicit TLRs (TLR2, TLR4, TLR6, and TLR9) seem to hold various jobs relying upon which arm of the inborn invulnerable reaction they draw in with, for instance, advancement of versatile reactions by working with antigen show in dendritic cells [93].

A few examinations have related normal hereditary variations (polymorphisms) in TLR qualities with vulnerability or inclination to foundational candidiasis or constant mucocutaneous candidiasis (CMC). These recollect polymorphisms for TLRI (R80T , N248S , and S602I) [94, 95] and TLR3 (L4I2F) [96, 97]. Polymorphisms in TLR4 (D299G) and TLR2 (D753Q) have moreover been perceived as possible frailty markers for basic candidiasis [98] yet these could not be approved in a greater report [95]. As of now, a large portion of the information accessible recommends a solid part for TLRs in antifungal protection however recognizing explicit jobs for each TLR has been over shadowed by repetitive signs instigated by other PRRs [94].

### 3.2 C-type lectin receptors

CLRs (C-type lectin receptors) are a diverse restriction protein family defined by the presence of an extracellular carb acknowledgment space (CRD) or a C-type lectin like area (CTLD) [99]. The job of CLRs in antifungal insusceptibility has been the subject of serious investigation as of late and a few key CLRs have now been shown to show basic capacities in Candida acknowledgment, take-up, and executing and furthermore add to the commencement and additionally tweak of the resistant reaction to organisms [46, 100, 101]. By and by, the key CTLs in Candida affirmation appear, apparently, to be dectin-I, dectin-2, and MR. CLRs signal through incitation of ITAM/ITIM (immunoreceptor tyrosine-based actuation/restraint theme) cytoplasmic areas (**Figure 3**). This can be done by using their own cytoplasmic area, as dectin-I does, or by using coreceptor cytoplasmic spaces, as DAPI2 (DNAX actuation protein of I2 kDa) and FcR (Fc receptor gamma chain) do, as dectin-2 does. The activation of numerous connections to those activated

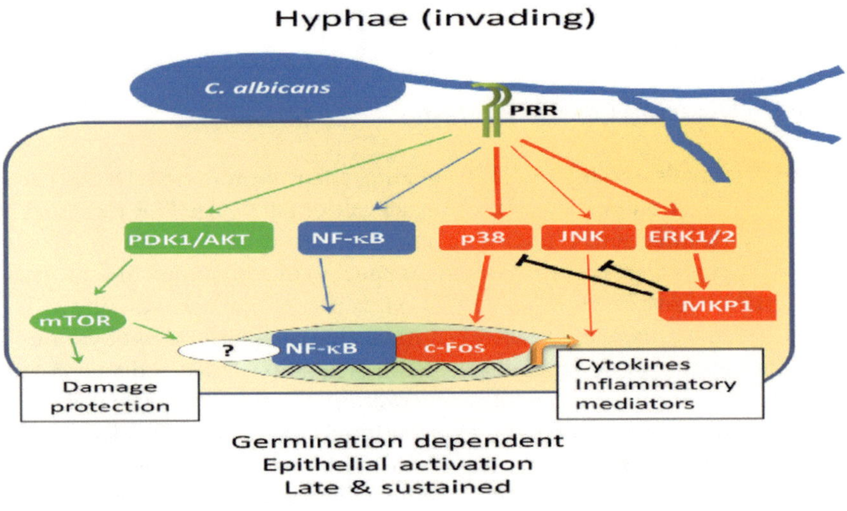


**Figure 3.**
*Signaling and damage pathways activated by* C. albicanshyphae. C. albicanshyphal *cells, when in sufficient quantities, are recognized by an unknown PRR mechanism that results in the activation of NF-κB, MAPK, and PI3K pathways. MAPK signaling via p38 and ERK1/2 appears to discriminate between yeast and hyphal cells. Activation of p38 by hyphae leads to activation of the c-Fos transcription factor, which, in conjunction with the p65/p50 NF-κB heterodimers and PI3K/AKT results in upregulation of cytokine and inflammatory mediator expression. Concurrently, activation of ERK1/2 signaling, results in stabilization of the MKP1 phosphatase, which deactivates p38 and JNK, hence acting as part of a negative feedback loop and preventing a potentially deleterious overreaction of theimmune system. Damage induced by hyphae appears to be mediated via JNK activation and prevented via the PI3K/AKT/mTor pathway.*

by TLRs, most notably Src family kinases including Src, Lyn, and Fyn, is triggered when CLRs are ligated. If we talk about dectin-I , it prompts initiation of spleen tyrosine kinase (SYK) and the downstream actuation of the CARD9/BclI0/MALTI (caspase enlistment space family/B cell CLL-lymphoma I0/mucosa related lymphoid tissue lymphoma movement quality I) flagging complex. Independent of the CLR pathways and connectors utilized, a definitive outcome is the enactment of comparative flagging pathways as those initiated by TLRs, overwhelmingly NF-$\kappa$B and MAPK, that are discussed below point.

### *3.2.1 Dectin-I*

Dectin-I, (also called CLEC7a) is that the main CLR known as taking part in a serious role in fungous recognition by the host system [102] and may be a sort II transmembrane macromolecule that belongs to a subgroup of CLRs referred to as natural killer (NK) receptor-like CLRs. The target ligands of dectin-I are $\beta$-I,3 glucan polymers, that comprise a serious part (~60%) of fungous cell walls. The intracellular region of dectin-I contains a changed ITAM motif containing one amino acid residue rather than the standard 2 (hence the terms hem-ITAM or hemi- ITAM). Activation of the dectin-I results in phosphorylation of this domain and phosphorylation of SYK and activation of the BclI0- CARD9-MALTI complicated as mentioned on top of. This results in activation of each the canonical and noncanonical NF-$\kappa$B pathways [103] further as nuclear issue of activated T cells (NFAT) pathway [104]. Dectin-I can even induce signaling via Raf-I in an exceedingly SYK -dependent fashion [103] and is related to phospholipase C and A2 activation [50]. one in all the most important functions of dectin-I binding seems to be the induction of bodily process [105]. However, a singular feature of dectin-I is its ability to be activated or suppressed by its target matter. to completely activate dectin-I, cells got to be exposed to insoluble $\beta$-glucan particles. Notably, exposure of dectin-I to soluble $\beta$-glucan seems to dam activation. This appears to ensue to the apparent form type a vegetative cell conjunction,"whereby phosphatases that usually suppress ITAM motifs are accumulated. This exclusion later permits the phosphorylation of the intracellular hem-ITAM motif [106], thereby sanctioning bodily process. Dectin-I has additionally been shown to synergise with each TLR2 and TLR4, leading to the induction of tumor necrosis factor (TNF), IL-I0, transforming growth factor (TGF) and dendritic cell maturation [107–109]. In view of the fact that the $\beta$ - I,3 glucan polymers that are the main components of the fungal cell wall, and a strong activation of the immune system, dectin-I plays an important role in inducing antifungal activity of the host. This may also explain why some of the mold surface structure of "the mask" -I.3 glucan from the immune system. For example, Histoplasma capsulatum, masks are $\beta$-I,3 glucan, with a low - $\alpha$-I,3 glucan [110] and it seems likely that the *C. albicans* hyphae of $\beta$-I,3 glucan has been covered over by layers of *N* - *O* - linked mannoproteins in order to prevent the discovery of the dectin-I. However, the yeast is in the form of *C. albicans*, while *N* - *O* - linked mannoproteins present in the underlying $\beta$ -glucan layer is exposed in the developing gut, which dectin-I in order to be recognized. Thus, it could be concluded that the most important role of dectin-I in the control of the yeast form of candidiasis (thrush). In addition, $\beta$ - glucan, which has been in the hyphal cell wall of *C. albicans*, it seems to be structurally different from the yeast $\beta$ - glucan [111] and, therefore, may not be immune to or understood by the dectin-I.

Although some studies have shown that the expression of dectin-I in the epithelial cells of the gastro- intestinal tract [112], and lung [113, 114], in oral epithelial cells express dectin-1 [115, 116]. What's interesting is that dectin-I expression appears to be reduced in the presence of live *C. albicans* cells [116], and it is not affected by the dectin-I ligands [115, 117]. This suggests that dectin-I is likely to play only a minor role in the detection of *C. albicans* epithelial cells. Studies carried out with the help of dectin-I knockout mice have provided mixed data sets for the *C. albicans* systemic infection models, to demonstrate these differences, [118] and increased mortality [119] depending on the study, the *C. albicans* strain used. On the one hand, it is the work for the dectin-I is supported by the consideration that CARD9 knockout mice are susceptible to the most important infection [120] and in patients with head-and-CARD9 immunity and are particularly vulnerable to both the lining and the main foundation candidiasis [121]. In addition, another study, it has been the study of the normal function of the genetic polymorphism in CARD9 (SI2N), to CADR9, especially candidiasis, it is recommended that the method of fixing of the $\beta$ - glucan may be excessive for the first invulnerability of *C. albicans* [122]. However, recent studies have shown the potential role of dectin-I in the maintenance of tissue health. Dectin-I−/− the mice showed greater severity of the disease, at least one more commonly, however, this weight can be reduced by the removal of fungal and bacterial flora in [123]. Histologically, extensive infestations of fungi have been recorded from the underlying tissue, which was not seen in wild-type mice. Clinical trial data have shown that a subgroup of patients with ulcerative colitis, especially in aggressive disease, and shows a common single-nucleotide polymorphisms (rs2078I78 in dectin-I, possibly indicating a requirement for functional dectin-I receptors, and to maintain, mucosal health, in a commensal state [123]. However, the role of dectin-I in the intramucosal infections, it is far from clear, as recent studies in mice have demonstrated that dectin-I does not play an important role in the control of gastro-intestinal colonization by *C. albicans* [124]. In particular, it is well known that, in humans, mutations in the stop codon (Tyr238X in dectin-I is associated with an increased risk of developing mucocutaneous fungal infections, with an increased colonization of the oral cavity and the gastrointestinal (gi) tract and vulvovaginal candidiasis (thrush) infection (RVVC) [125, 126]. In another case, we obtained that the dectin-I polymorphism (I223S) was associated with oropharyngeal candidiasis (OPC) the susceptibility of West Africa, a group of HIV-positive patients [127]. That is why, even though the great one, the precise role of dectin-I in the susceptibility to candida infection is still unclear and requires further investigation.

#### *3.2.2 Dectin-2*

Dectin-2 (otherwise called CLEC6a) is a sort II transmembrane protein however is enacted contrastingly to dectin-I. Dectin-2 comes up short on an intracellular flagging area [128] and requirements to dimerise with FcR$\gamma$, which has an intracellular flagging space, to send a sign [69]. In myeloid cells and fiery monocytes, dectin-2 perceives high mannose structures that are normal to numerous parasites and ties to hyphae with higher proclivity than to yeast [129, 130]. This may clarify why dectin-2 inadequate mice are helpless to *C. albicans* contamination be that as it may, strangely, not *C. neoformans* [130, 131]. Dectin-2 may moreover recognize $\alpha$-mannosyl linkages [132]. Dectin-2 may activate a number of cytokines and chemokines via NF-B, MAPK, SYK, CARD9-BclI0-MaltI, and PKC, as well as initiate the NLRP3 (NOD-like receptor family, pyrin region containing 3) inflammasome and a respiratory burst [69, 133]. Furthermore, dectin-2 may have a role in protecting against *C. glabrata* illnesses, since dectin-2/lacking mice were more susceptible to *C. glabrata* infections, indicating a poor transmittable choice in kidneys [134].

#### *3.2.3 Dectin-3*

Dectin-3 (additionally called CLECsf8, MCL, or CLEC4d) was as of late distinguished and seems to shape heterodimers with dectin-2 to perceive $\alpha$-mannans on the outside of *C. albicans* hyphae, prompting NF-$\kappa$B enactment [135]. Strikingly, dectin-3−/− mice were exceptionally helpless to *C. albicans* disease. Contrasted and their particular homodimers, dectin-2/3 heterodimers bound $\alpha$-mannans all the more viably, prompting strong incendiary reactions. This recommends that distinctive CLRs may shape an assortment of hetero and homodimers that may give diverse affectability and variety to have cells to identify different contagious contaminations.

#### *3.2.4 DC-SIGN*

DC-SIGN (otherwise called CD209) is another sort II transmembrane receptor that is communicated dominatingly on dendritic cells and macrophages. Dendritic Cell-Specific Intercellular adhesion molecule-3-Grabbing Non-integrin) also known as CD209 (Cluster of Differentiation 209) is a protein which in humans is encoded by the CD209 gene. DC-SIGN is a C-type lectin receptor present on the surface of both macrophages and dendritic cells Nonetheless, the part of DC-SIGN in antifungal invulnerability is muddled [101], in spite of the fact that DC-SIGN seems to perceive high (*N*-connected) mannose containing glycoproteins and actuate IL-6 creation [71, 136]. Albeit the part of DC-SIGN in the endocytosis and take-up of microbes to advance antigen show is all around recorded [136, 137], its job in phagocytosis is sketchy [71, 136].

### 3.3 Mannose receptor

The MR (or called CD206) is a prototypical kind I transmembrane protein that is transcendently communicated on macrophage and dendritic cells. MR receptor ties a few starch particles, including extended *N*-connected mannans, N-acetylglucosamine, glucose, and fucose [138]. Thus, MR can perceive numerous contagious, bacterial, and viral pathogens. MR needs regular intracellular flagging spaces despite the fact that ligation actually prompts an assortment of cell reactions, including signal pathway acceptance, phagocytosis, advancement of antigen show to T cells, and cytokine discharge [63, 136–140]. For instance, the MR is enlisted to the phagosome after *C. albicans* ingestion and actuates intracellular flagging and cytokine creation [141]. MR may likewise be needed for the enlistment of defensive ThI7 reactions in *C. albicans* contamination [140] however may repress cytokine creation because of different organisms, for instance, Pneumocystis carinii [142]. Remarkably, MR inadequacy does not seem to present helplessness to *C. albicans* foundational disease [143] as it does to *C. neoformans* [144], albeit minor changes in parasitic weights can be noticed [143]. In oral epithelial cells, MR impeding does not modify the discharge of IL-6, IL-8, and GM-CSF upon incitement with Candida cell divider parts [117]. As of now, there is no conclusive part for MR in mucosal antifungal host safeguards.

### 3.4 Nod-like receptors (NLRs) and inflammasomes

NLRs are a group of intracellular PRRs portrayed by leucine rich rehashes and a nucleotide-restricting area that identify PAMPs present in the cell cytoplasm. Like TLRs and CTLs, NLRs perceive microbial items yet they additionally perceive have determined threat signals or alarmins [145]. There are now 23 human NLRs and 34 mouse NLRs identified [146]. Inflammasomes are huge multimeric protein structures framed by NLRs and two distinct proteins, ASC (apoptosis-related spot like protein containing a CARD) and procaspase-I (procysteine-subordinate aspartate-coordinated protease I). The inflammasome's main function is to convert procaspase-I to dynamic caspase-I, which causes young cells that are friendly to IL-I and supportive of IL- I8 to produce IL-I and IL-I8 [147]. Despite the fact that *C. albicans* is not recognized by NLRCI (NLR family CARD space containing protein I) or NLRC2 [148], it is known to activate inflammasomes fusing NLRP3 (NACHT , LRR, and PYD spaces containing protein 3) [149] and NLRC4 [150], resulting in the production of IL-I.

Surprisingly, NLRP3 is strongly expressed in nonkeratinizing epithelia, such as the oral cavity and throat [151], suggesting a possible role for NLRP3 in parasitic recognition in oral epithelial cells, which is supported by studies showing increased IL- I and IL- I8 levels in response to *C. albicans* stimulation [115, 152–155]. Mice missing NLRP3 appear to be susceptible to candidiasis [156], but mice lacking IL-I receptor type I (IL- IRI), IL-I8, or caspase-I exhibit distinct contagious contamination helplessness profiles [157]. Strikingly, IL-I$\beta$ (and IL-I$\alpha$) lacking mice show expanded mortality during scattered candidiasis [158]. Late reports have likewise recognized a significant part for NLRP3 along with TLR2 and dectin-I in forestalling dispersal of *C. albicans* in a murine model of oral contamination [159]. Steady with a part for NLRP3 inmucosal security [160], deficient NLRP3 actuation expands *C. albicans* colonization in the gut and fuels Crohn's illness [161], and a length polymorphism in intron 4 of the quality (CIASI) that codes for NLRP3 inclines patients to RVVC [162]. Nevertheless, the full degree of the practical jobs for NLRs and inflammasomes in antifungal host safeguards is as yet not completely comprehended.

## 4. Protein involves in pathogenesis

### 4.1 Mincle

Mincle (also known as CLEC4e or CLECsf9) is a type II transmembrane protein that transmits its signal after dimerizing with the FcR connector protein [128]. Macrophages, monocytes, neutrophils, myeloid dendritic cells, and certain B cell subsets all communicate mincle, while plasmacytoid dendritic cells, T cells, and NK cells do not [133]. Mincle binds -mannans-containing starch structures [143, 163] and detects *C. albicans* [70, 164, 165], Malassezia spp. [163], and Fonsecaea pedrosoi, the chromoblastomycosis causative pathogen [166]. As with dectin-2, mincle is not believed to be needed for phagocytosis [70] yet adds to the acceptance of cytokines and chemokines by means of NF-$\kappa$B, MAPK, SYK, CARD9-BclI0-MatIt, and PKC$\delta$ [133, 163]. In spite of the fact that mincle-incited reactions have all the earmarks of being MyD88 autonomous, mincle may synergise with TLRs to instigate fiery cytokines and the respiratory burst [167].

### 4.2 Soluble proteins in Candida recognition

The supplement course assumes a significant part in have protection against parasitic microorganisms and is quickly enacted in light of host attack by Candida [168–170]. Candida actuates each of the three known pathways (old style, elective, and mannose-restricting lectin (MBL)) with nobody clear pathway overwhelming the reaction [171]. Given that the Candida cell surface is covered with a bounty of manno proteins, it is not astonishing that Candida microorganisms are viable at actuating the MBL pathway, which seems significant for opsonisation, phagocytosis, and other supplement capacities [172, 173]. The connection between enacted C3b and the supplement receptor CR3 is generally needed for the uptake of Candida cells by phagocytes [174]. *C. albicans* cell divider proteins (e.g., GpmI, PraI, and Gpd2) can possibly tie supplement segments, for example, Factor H, FHL-I, C4BP and plasminogen from human plasma that meddle with phagocytic opsonisation and take-up [168, 170, 175–180]. For example, restricting of Pra1 to factor H and FHL-1 most likely includes an avoidance methodology including the hindrance of C3 cleavage into opsonic and anaphylatoxic parts, in this manner forestalling acknowledgment and take-up by phagocytes [181]. C5 is likewise significant in Candida diseases since mice that need practical C5 quality duplicates are vulnerable to obtrusive foundational contaminations [182–185]. C5 insufficiency is related with expanded creation of proinflammatory cytokines (TNF$\alpha$ and IL-6) and fast parasitic replication in organs that can prompt cardiovascular disappointment [186, 187]. Sanctioning of C5 prompts the improvement of C5b, which consequently triggers the plan of the film attack complex (MAC). Despite the fact that affidavit of MAC on the outside of *C. albicans* does not bring about fungicidal movement, presumably because of the thickness of the parasitic cell divider, it might work with the incitement of phagocytes and ensuing arrival of terminal supplement segments from these phones. Curiously, as no impact on irritation is recognized in C3 insufficient mice, this may recommend a generally C3-free preparing of C5 in foundational *C. albicans* disease [188]. After phagocytosis, the oxidative burst is set off which prompts contagious executing, a cycle that can be hindered with monoclonal antibodies to forestall C3b-CR3 associations. C3b-CR3 contact also appears to be crucial for lymphocyte hyphal formation and cytokine production [189]. MBL has also been linked to the inhibition of Candida development [190] and the enhancement of TNF release from Candida-infected monocytes [191]. C3a, an anaphylatoxin released by C3 during supplement enactment, may have direct antifungal activity independent of its chemotactic effect [192]. These findings suggest that complement activation is critical in the host's defense against *C. albicans* infections. The reader is directed to the following reviews [168, 170] for further in-depth information on the involvement of complement in Candida infections.

## 5. Cellular responses to Candida

### 5.1 Neutrophils

Neutrophils are a key effector cell in intrinsic insusceptibility, and they play a dual role in antifungal responses. First, they phagocytose and destroy contaminated Candida cells (below), and then, via cross communication with epithelial cells, they indirectly assist in mucosal protection (tended to above). TLRs and CTLs help neutrophils phagocytose nonopsonized Candida, while CR3 and the Fc receptor (FcR) help them phagocytose opsonized Candida [193]. Once phagocytosed, Candida is killed both inside and outside the cell through oxidative and nitrosative mechanisms, but fungicidal movement varies across Candida species [194, 195]. Preformed cytoplasmic granules interweave with the phagosome intracellularly, although unlike macrophages, no substantial pH changes occur [196]. Antimicrobial proteins found in neutrophil granules include defensins, lactoferrin, lysozyme, myeloperoxidase, and elastase [197], all of which can be transported into the extracellular environment. Candida's phagocytic execution requires oxidative processes. During the oxidative burst, neutrophils create reactive oxygen species (ROS), which needs the NADPH oxidase catalyst complex to assemble in the cytoplasmic and phagosomal film [198]. First, the superoxide extremist is formed, which is subsequently dismutated to hydrogen peroxide, an oxidative and harmful particle [199].

Then, myeloperoxidase uses hydrogen peroxide to create hypochlorous acid, which is moreover an exceptionally oxidative particle that responds with natural amines to frame chloramines that have further antimicrobial stuffs [193, 200]. Candida's phagocytic execution is further aided by reactive nitrogen species (RNS) [193]. When neutrophils are activated, they produce nitric oxide (NO) from arginine and oxygen via an enzyme called inducible nitric oxide synthase (iNOS). NO is extremely sensitive, and it is converted to peroxynitrite, which is then reduced to nitrogen dioxide and a hydroxyl radical. Because iNOS is restricted to the intracellular compartment, RNS production is restricted to the intracellular compartment [199]. The creation of neutrophil extracellular catches (NETs) [201, 202], which are formed during a unique sequence of neutrophil cell death known as NETosis, is another more recently found way of Candida executing. Similar to serine proteases, antimicrobial peptides (e.g., calprotectin), and other microbicidal chemicals, the neutrophil "explodes," unleashing a snare of chromatin fibrils coated with the neutrophil's material. Candida spp. are well-versed in surviving the oxidative, nitrosative, osmotic, and restorative nerves encountered during interactions with neutrophils. Because of the weights, many cycles, features, and proteins are altered within the organism. These include upregulation of transporters (e.g., oligopeptide, ammonium, and iron), use of alternative carbon and nitrogen sources and metabolic cycles (e.g., glycolysis, glyoxylate, unsaturated fat, and amino destructive), and detoxification of neutrophil oxidative/nitrosative butchering instruments. (e.g., catalase, superoxide dismutases, and nitric oxide dioxygenase). In any event, these nuances are beyond the scope of this examination, and the reader is directed to a later examination that focuses on the Candida reaction to neutrophils [193, 203].

## 5.2 Macrophages

Macrophages can function as phagocytic cells as well as antigen-presenting cells capable of activating T lymphocytes. Upon activation, macrophages divide into two phenotypically and functionally distinct subsets, M1 and M2, based on the cytokine milieu in which they are initiated [204–206]. The M1 total is derived from receptiveness to the T colleague (Th)1 cytokine IFN, whereas the M2 total is derived from receptiveness to Th2 cytokines, IL-4 and IL- 13. M1 macrophages are microbic and proinflammatory, whereas M2 macrophages are involved in wound healing and extracellular network upgradation. Macrophages, like neutrophils, see and phagocytoze nonopsonised Candida via TLRs and CTLs, and opsonised Candida via CR3 and FcR [193, 207]. Nonetheless, macrophage phagosome formation differs from neutrophil phagosome development in that macrophage phagosomes follow the endocytic development route and grow into phagolysosomes with a distinctive acidic pH that promotes compound activity, such as cathepsin D [208]. M1 macrophages use both oxidative and nitrosative executing components (as seen above for neutrophils), but they also use the RNS, NO, to directly kill phagocytosed Candida via the translocation of iNOS. TNF and the chemokines CXCL9 and CXCL10 are also released by M1 macrophages [209]. These chemokines act as ligands for the CXCR3 receptor, which is found on Th1 cells and NK cells, attracting resistant cells to contamination sites.

M2 macrophages, then again, advance contagious ingenuity inside the macrophage, giving an instrument to invulnerable avoidance. M2 macrophages additionally express more significant levels of MR (CD206) bringing about expanded phagocytosis of Candida [210]. Correspondingly, the arginase-1 (Arg1) quality is additionally expanded in articulation, which rivals iNOS for a similar substrate (arginine), consequently diminishing NO levels [211]. This is additionally exacerbated by decreased degrees of TNF$\alpha$ creation in M2 macrophages. In light of this, macrophages anticipate playing an important role in Candida protection, but this is contingent on the Candida strain assisting the macrophage [212]. Candida spp., like neutrophils, are believed to rely on relative adaptations to survive in macrophages. *C. albicans* and *C. glabrata* have been shown to alter metabolic requirements by using alternative carbon sources, upregulating impetuses for gluconeogenesis, glyoxylate cycle, and -oxidation of unsaturated lipids, and downregulating protein synthesis and glycolysis [193, 207]. This combines the formation of catalase and superoxide dismutases for extracellular ROS detoxification [213] and the outflow of flavohemoglobin impetuses for intracellular RNS butchering [214]. Concerning *C. albicans*, intracellular dealing additionally seems unusual and the growth may repress both lysosomal fermentation and NO delivery [215]. For additional subtleties the peruser is guided to ongoing surveys that emphasis on the Candida reaction to macrophages [207].

Besides these receptors molecules, actively participated proteins and cellular mechanism system there is a lot of others factors in these mechanisms are linked like adhesins and invasins, biofilm formation, contact sensing and thigmotropism, secreted hydrolases, pH-sensing and its regulation, environment and metabolic adaptation, small HSPs, metal acquisition. So, for a complete understanding these factors also play significant role in pathogenicity mechanism of *C. albicans*.

## 6. Conclusion

This chapter has discussed the pathogenicity mechanism along with host and cellular responses in Candida species. Host reactions to Candida are profoundly assorted because of the assortment of contagious PAMPs and antigens perceived by various safe cells at different disease destinations. Many inquiries have been conducted on this important topic, particularly with *C. albicans*, and thus we have obtained a much improved understanding of the appropriate structures of the PAMPs & PRRs. Still, further analysis is needed in order to attain insight into the complex communication between PAMPs and the corresponding receptors. Definitely, co-stimulation via multiple PAMP–PRR interactions may increase together the sensitivity as well as the specificity of the immune recognition process.

## Author details

Snigdha Pattnaik*, Laxmidhar Maharana and Manoj Sethi
Siksha 'O' Anusandhan, India

*Address all correspondence to: snigdhapattnaik@soa.ac.in

## References

[1] Vincent JL, Rello J, Marshall J, Silva E, Anzueto A, Martin CD et al (2009) EPIC II Group of Investigators. International study of the prevalence and outcomes of infection in intensive care units. JAMA 302:2323-2329.

[2] Ingham CJ, Boonstra S, Levels S, de Lange M, Meis JF, Schneeberger PM (2012) Rapid susceptibility testing and microcolony analysis of Candida spp. cultured and imaged on porous aluminium oxide. PLoS ONE 7:e33818.

[3] Correia A, Sampaio P, Vilanova M, Pais C (2015) *Candida albicans*: clinical relevance, pathogenesis, and host immunity. In: Sing SK (ed) Human emerging and re-emerging infections: viral and parasitic infections, vol 1. John Wiley and Sons, New Jersey, pp 926-952.

[4] Limon JJ, Skalski JH, Underhill DM (2017) Commensal fungi in health and disease. Cell Host Microbes 22:156-165.

[5] De Rosa FG, Garazzino S, Pasero DC, Peri GD (2009) Invasive candidiasis and candidemia: new guidelines. Minerva Anaestesiologica 75:453-458.

[6] Negri M, Faria M, Guilhermetti E, Alves A, Paula C, Svidzinski T (2010) Hemolytic activity and production of germ tubes related to pathogenic potential of clinical isolates of *Candida albicans*. J Basic Appl Pharm. 31:89-93.

[7] Silva S, Negri M, Henriques M, Oliveira R, Williams DW, Azeredo J (2011) Adherence and biofilm formation of non- *Candida albicans* Candida species. Trends Microbio 19:241-247.

[8] Wisplinghoff H, Seifert H, Tallent SM, Bischoff T, Wenzel RP, Edmond MB (2003a) Nosocomial bloodstream infections in pediatric patients in United States hospitals: epidemiology, clinical features and susceptibilities. Pediatr Infect Dis J 22:686-691.

[9] Bongomin F, Gago S, Oladele R, Denning D (2017) Global and multi-national prevalence of fungal diseases-estimate precision. J Fungi 3:57.

[10] Dadar M, Tiwari R, Karthik K, Chakraborty S, Shahali Y, Dhama K (2018) *Candida albicans*-biology, molecular characterization, pathogenicity, and advances in diagnosis and control-an update. Microb Pathog 117:128-138.

[11] Caceres DH, Forsberg K, Welsh RM, Sexton DJ, Lockhart SR, Jackson BR et al (2019) Candida auris: a review of recommendations for detection and control in health care settings. J Fungi 5:111.

[12] Aslam B,Wang W, Arshad MI, Khurshid M,Muzammil S, Rasool MH et al (2018) Antibiotic resistance: a rundown of a global crisis. Infect Drug Resist 11:1645-1658.

[13] Kornitzer D (2019) Regulation of *Candida albicans* hyphal morphogenesis by endogenous signals. J Fungi 5:21.

[14] Wisplinghoff H, Elobers J, Geurtz L, Stefanik D, Major Y, Edmond MB et al (2014) Nosocomial bloodstream infections due to Candida spp. in the USA: species distribution, clinical features and antifungal susceptibilities. Int J Antimicrob Agents 43:78-81

[15] Perez JC, Johnson AD (2013) Regulatory circuits that enable proliferation of the fungus *Candida albicans* in a mammalian host. PLoS Pathogen 9(12):e1003780.

[16] Jacobsen ID, Hube B (2017) *Candida albicans* morphology: still in focus. Expert Rev Anti Infect Ther 15:327-330.

[17] AokiW, Kitahara N,Miura N, Morisaka H, Yamamoto Y, Kuroda K et al (2011) Comprehensive characterization of secreted aspartic proteases encoded by a virulence gene family in *Candida albicans*. J BioChem 150:431-438.

[18] Seman BG, Moore JL, Scherer AK, Blair BA, Manandhar S, Jones JM et al (2018) Yeast and filaments have specialized, independent activities in a zebrafish model of *Candida albicans* infection. Infect Immun 86:e00415–e00418.

[19] Desai JV, Cheng S, Ying T, NguyenMH, Clancy CJ, Lanni F et al (2015) Coordination of *Candida albicans* invasion and infection functions by phosphoglycerol phosphatase Rhr2. Pathogens 4: 573-589.

[20] Kadosh D (2017) Morphogenesis in *C. albicans*. In: Prasad R (ed) *Candida albicans*: Cell Mol Biol. Springer, Cham

[21] Han TL, Cannon RD, Villas-Boas SG (2011) The metabolic basis of *Candida albicans* morphogenesis and quorum sensing. Fungal Genet Biol 48:747-763.

[22] Monge RA, Román E, Nombela C, Pla J (2006) The MAP kinase signal transduction network in *Candida albicans*. Microbiology 152:905-912.

[23] Gong Y, Li T, Yu C, Sun S (2017) *Candida albicans* heat shock proteins and Hsps-associated signaling pathways as potential antifungal targets. Front Cell Infect Microbiol 7:520.

[24] Smith DA, Nicholls S,Morgan BA, Brown AJP, Quinn JA (2004) Conserved stress-activated protein kinase regulates a core stress response in the human pathogen *Candida albicans*.Mol Biol Cell 15:4179-4190.

[25] Hogan D, Sundrom P (2009) The Ras/Camp/PKA signaling pathways and virulence in *Candida albicans*. Future Microbiol 4: 1263-1270.

[26] Lin C-J, Wu C-Y, Yu S-J, Chen Y-L (2018) Protein kinase A governs growth and virulence in *Candida tropicalis*. Virulence 9(1):331-347.

[27] Inglis DO, Sherlock G (2013) Ras signaling gets fine-tuned: regulation of multiple pathogenic traits of *Candida albicans*. Eukaryot Cell 12:1316-1325.

[28] Lin CJ, Chen YL (2018) Conserved and divergent functions of the cAMP/PKA signaling pathway in *Candida albicans* and *Candida tropicalis*. J Fungi 4:68.

[29] DavisDA (2009) Howhuman pathogenic fungi sense and adapt to pH: the link to virulence. Curr Opin Microbiol 12:365-370.

[30] Brown A, Haynes K, Gow N, Quinn J (2012) Stress responses in Candida, 2nd edn. ASM Press, Washington, D.C., pp 225-242.

[31] Zhou Y, LiaoM, Zhu C, Hu Y, Tong T, Peng X et al (2018) ERG3 and ERG11 genes are critical for the pathogenesis of *Candida albicans* during the oral mucosal infection. Int J Oral Sci 10:9.

[32] de Oliveira SGC, Vasconcelos CC, Lopes AJO, de Sousa Cartagenes MDS, Filho AKDB, do Nascimento FRF et al (2018) Candida infections and therapeutic strategies: mechanisms of action for traditional and alternative agents. Front Microbiol 9:1351.

[33] Dantas SA, Lee KK, Raziunaite I, Schaefer K, Wagener J, Yadav B et al (2016) Cell biology of *Candida albicans*-host interactions. Curr Opin Microbiol 34:111-118.

[34] Schonherr FA, Sparber F, Kirchner FR, Guiducci E, Trautweinweidner K, Gladiator A et al

(2017) The interspecies diversity of *C. albicans* triggers qualitatively and temporally distinct host responses that determine the balance between commensalism and pathogenicity. Mucosal Immunol 10:1335-1350.

[35] Braunsdorf C, LeibundGut-Landmann S (2018) Modulation of the fungal-host interaction by the intra-species diversity of *C. albicans*. Pathogens 7:11.

[36] Reedy JL, Filler SG, Heitman J (2010) Elucidating the *Candida albicans* calcineurin signaling cascade controlling stress response and virulence. Fungal Genet Biol 47:107.

[37] Liu S, Liu W (2015) Components of the canclium-calcinerium signaling pathways in fungal cells and their potential as antifungal targets. Eukaryot Cell 14:4.

[38] Yu Q, Jia C, Dong Y, Zhang B, Xiao C, Chen Y et al (2015) *Candida albicans* autophagy, no longer a bystander: its role in tolerance to ER stress-related antifungal drugs. Fungal Genet Biol 81:238-249.

[39] Shang-Jie Y, Ya-Lin C, Ying-Lie C (2015) Calcineurin signaling: lessons from Candida species. FEMS Microbiol 15:4.

[40] Wang L, Lin X (2012) Morphogenesis in fungal pathogenesis: shape, size and surface. PLoS Pathog 8:e1003027.

[41] Kim S, Nguyen QB,WolyniakMJ, Frechette G, Lehman CR, Fox BK et al (2018) Release of transcriptional repression through the HCR promoter region confers uniform expression of HWP1 on surfaces of *Candida albicans* germ tubes. PLoS ONE 13: e0192260.

[42] Sharma J, Rosiana S, Razzaq I, Shapiro RS (2019) Linking cellular morphogenesis with antifungal treatment and susceptibility in Candida pathogens. J Fungi 5:17.

[43] Sun JN, Solis NV, Phan QT, Bajwa JS, Kashlera H, Thompson A et al (2010) Host cell invasion and virulence mediated by *Candida albicans* Ssai. PLos Pathog 6:e1001181.

[44] Wächtler B,Wilson D, Haedicke K, Dalle F, Hube B (2011) From attachment to damage: defined genes of *Candida albicans* mediate adhesion, invasion and damage during interaction with oral epithelial cells. PLoS One 6:e17046.

[45] C. A. Janeway Jr. and R. Medzhitov, "Innate immune recognition," *Annual Review of Immunology*, vol. 20, pp. 197-216, 2002.

[46] M. G. Netea, G. D. Brown, B. J. Kullberg, and N. A. R. Gow, "An integrated model of the recognition of *Candida albicans* by the innate immune system,"*Nature ReviewsMicrobiology*, vol. 6, no. 1, pp. 67-78, 2008.

[47] J. R. Naglik and D. Moyes, "Epithelial cell innate response to *Candida albicans*.," *Advances in dental research*, vol. 23, no. 1, pp. 50-55, 2011.

[48] A. Roeder, C. J. Kirschning, R. A. Rupec, M. Schaller,G.Weindl, and H. C. Korting, "Toll-like receptors as key mediators in innate antifungal immunity," *Medical Mycology*, vol. 42, no. 6, pp. 485-498, 2004.

[49] G. Weindl, J. Wagener, and M. Schaller, "Epithelial cells and innate antifungal defense," *Journal of Dental Research*, vol. 89, no. 7, pp. 666-675, 2010.

[50] A. Plato, J. A. Willment, and G. D. Brown, "C-Type lectinlike receptors of the dectin-1 cluster: Ligands and signalling pathways," *International Reviews of Immunology*, vol. 32, no. 2, pp. 134-156, 2013.

[51] D. J. Philpott, M. T. Sorbara, S. J. Robertson, K. Croitoru, and S. E. Girardin, "NOD proteins: regulators of inflammation in health and disease," *Nature Reviews Immunology*, vol. 14, pp. 9-23, 2014.

[52] C. Bourgeois and K. Kuchler, "Fungal pathogens—a sweet and sour treat for toll-like receptors," *Frontiers in Cellular and Infection Microbiology*, vol. 2, article 142, 2012.

[53] S. Akira, "Mammalian Toll-like receptors," *Current Opinion in Immunology*, vol. 15, no. 1, pp. 5-11, 2003.

[54] K. Takeda, T. Kaisho, and S. Akira, "Toll-like receptors," *Annual Review of Immunology*, vol. 21, pp. 335-376, 2003.

[55] S. Akira and K. Takeda, "Toll-like receptor signalling," *Nature Reviews Immunology*, vol. 4, no. 7, pp. 499-511, 2004.

[56] K. Takeda and S. Akira, "TLR signaling pathways," *Seminars in Immunology*, vol. 16, no. 1, pp. 3-9, 2004.

[57] T. Kawai and S. Akira, "Pathogen recognition with Toll-like receptors," *Current Opinion in Immunology*, vol. 17, no. 4, pp. 338-344, 2005.

[58] T. Kawai and S. Akira, "TLR signaling," *Seminars in Immunology*, vol. 19, no. 1, pp. 24-32, 2007.

[59] B. Lemaitre, E. Nicolas, L. Michaut, J. Reichhart, and J. A. Hoffmann, "The dorsoventral regulatory gene cassette spatzle/Toll/Cactus controls the potent antifungal response in Drosophila adults," *Cell*, vol. 86, no. 6, pp. 973-983, 1996.

[60] T. Jouault, S. Ibata-Ombetta, O. Takeuchi et al., "*Candida albicans* phospholipomannan is sensed through toll-like receptors," *Journal of Infectious Diseases*, vol. 188, no. 1, pp. 165-172, 2003.

[61] G. D. Brown, P. R. Taylor, D.M. Reid et al., "Dectin-1 is a major $\beta$-glucan receptor on macrophages," *Journal of Experimental Medicine*, vol. 196, no. 3, pp. 407-412, 2002.

[62] K. M. Dennehy, J. A. Willment, D. L. Williams, and G. D. Brown, "Reciprocal regulation of IL-23 and IL-12 following coactivation of dectin-1 and TLR signaling pathways," *European Journal of Immunology*, vol. 39, no. 5, pp. 1379-1386, 2009.

[63] M. G. Netea, N. A. R. Gow, C. A. Munro et al., "Immune sensing of *Candida albicans* requires cooperative recognition of mannans and glucans by lectin and Toll-like receptors," *Journal of Clinical Investigation*, vol. 116, no. 6, pp. 1642-1650, 2006.

[64] G. Ferwerda, F. Meyer-Wentrup, B. Kullberg, M. G. Netea, and G. J. Adema, "Dectin-1 synergizes with TLR2 and TLR4 for cytokine production in human primary monocytes and macrophages," *Cellular Microbiology*, vol. 10, no. 10, pp. 2058– 2066, 2008.

[65] A.Ozinsky, D.M.Underhill, J.D. Fontenot et al., "The repertoire for pattern recognition of pathogens by the innate immune system is defined by cooperation between Toll-like receptors," *Proceedings of the National Academy of Sciences of the United States of America*, vol. 97, no. 25, pp. 13766-13771, 2000.

[66] M. G. Netea, F. Van De Veerdonk, I. Verschueren, J.W. M. Van DerMeer, andB. J. Kullberg, "RoleofTLR1 andTLR6 in thehost defense against disseminated candidiasis," *FEMS Immunology and Medical Microbiology*, vol. 52, no. 1, pp. 118-123, 2008.

[67] S. P. Smeekens, F. L. van de Veerdonk, J. W. M. van der Meer, B. J. Kullberg, L. A. B. Joosten, andM.

G.Netea, "The Candida Th17 response is dependent on mannanand $\beta$-glucan-induced prostaglandin E2," *International Immunology*, vol. 22, no. 11, pp. 889-895, 2010.

[68] T. Jouault, M. El Abed-El Behi, M. Mart'ınez-Esparza et al., "Specific recognition of *Candida albicans* by macrophages requires galectin-3 to discriminate *Saccharomyces cerevisiae* and needs association with TLR2 for signaling," *Journal of Immunology*, vol. 177, no. 7, pp. 4679-4687, 2006.

[69] K. Sato, X. Yang, T. Yudate et al., "Dectin-2 is a pattern recognition receptor for fungi that couples with the Fc receptor $\gamma$ chain to induce innate immune responses," *The Journal of Biological Chemistry*, vol. 281, no. 50, pp. 38854-38866, 2006.

[70] C. A. Wells, J. A. Salvage-Jones, X. Li et al., "The macrophageinducible C-type lectin, mincle, is an essential component of the innate immune response to *Candida albicans*," *Journal of Immunology*, vol. 180, no. 11, pp. 7404-7413, 2008.

[71] A. Cambi, K. Gijzen, I. J. M. de Vries et al., "The C-type lectin DC-SIGN (CD209) is an antigen-uptake receptor for *Candida albicans* on dendritic cells," *European Journal of Immunology*, vol. 33, no. 2, pp. 532-538, 2003.

[72] P.R.Taylor,G.D. Brown, J.Herre,D. L.Williams, J. A.Willment, and S.Gordon, "TheRole of SIGNR1 and the $\beta$-GlucanReceptor (Dectin-1) in the Nonopsonic Recognition of Yeast by Specific Macrophages," *Journal of Immunology*, vol. 172, no. 2, pp. 1157-1162, 2004.

[73] M. G. Netea and B. J. Kullberg, "Epithelial sensing of fungal invasion," *Cell Host andMicrobe*, vol. 8, no. 3, pp. 219-220, 2010.

[74] A. Esteban, M.W. Popp, V. K. Vyas, K. Strijbis,H. L. Ploegh, and G. R. Fink, "Fungal recognition is mediated by the association of dectin-1 and galectin-3 in macrophages," *Proceedings of the National Academy of Sciences of the United States of America*, vol. 108, no. 34, pp. 14270-14275, 2011.

[75] *L. Romani*, "Immunity to fungal infections," *Nature Reviews Immunology*, vol. 11, no. 4, pp. 275-288, 2011.

[76] L. B. Ivashkiv, "A signal-switch hypothesis for cross-regulation of cytokine and TLR signalling pathways," *Nature Reviews Immunology*, vol. 8, no. 10, pp. 816-822, 2008.

[77] S. V. Tsoni and G. D. Brown, "$\beta$-Glucans and dectin-1," *Annals of the New York Academy of Sciences*, vol. 1143, pp. 45-60, 2008.

[78] E. F. Kenny and L. A. J. O'Neill, "Signalling adaptors used by Toll-like receptors: an update," *Cytokine*, vol. 43, no. 3, pp. 342-349, 2008.

[79] L. A. J. O'Neill, "The interleukin-1 receptor/Toll-like receptor superfamily: 10 Years of progress," *Immunological Reviews*, vol. 226, no. 1, pp. 10-18, 2008.

[80] E. Villam'on, D. Gozalbo, P. Roig et al., "Myeloid differentiation factor 88 (MyD88) is required formurine resistance to *Candida albicans* and is critically involved in Candida-induced production of cytokines," *European Cytokine Network*, vol. 15, no. 3, pp. 263-271, 2004.

[81] C. Bourgeois,O.Majer, I. E. Frohner, L. Tierney, and K. Kuchler, "Fungal attacks on mammalian hosts: pathogen elimination requires sensing and tasting," *Current Opinion in Microbiology*, vol. 13, no. 4, pp. 401-408, 2010.

[82] S. Bellocchio, C. Montagnoli, S. Bozza et al., "The contribution of the Toll-like/IL-1 receptor superfamily to innate and adaptive immunity to fungal

pathogens in vivo," *The Journal of Immunology*, vol. 172, no. 5, pp. 3059-3069, 2004.

[83] M. L. Gil andD. Gozalbo, "Role of toll-like receptors insystemic *Candida albicans* infections," *Frontiers in Bioscience*, vol. 14, no. 2, pp. 570-582, 2009.

[84] M. G. Netea, C. A. A. van der Graaf, A. G. Vonk, I.Verschueren, J. W. M. Van der Meet, and B. J. Kullberg, "The role of tolllike receptor (TLR) 2 and TLR4 in the host defense against disseminated candidiasis," *Journal of Infectious Diseases*, vol. 185, no. 10, pp. 1483-1489, 2002.

[85] M.G.Netea, R. Sutmuller, C.Hermann et al., "Toll-like receptor 2 suppresses immunity against *Candida albicans* through induction of IL-10 and regulatory T cells," *Journal of Immunology*, vol.172, no. 6, pp. 3712-3718, 2004.

[86] M. G. Netea, N. A. R. Gow, L. A. B. Joosten, I. Verschueren, J. W. M. Van Der Meer, and B. J. Kullberg, "Variable recognition of *Candida albicans* strains by TLR4 and lectin recognition receptors," *Medical Mycology*, vol. 48, no. 7, pp. 897-903, 2010.

[87] E. Villam'on, D. Gozalbo, P. Roig et al., "Toll-like receptor 2 is dispensable for acquired host immune resistance to *Candida albicans* in a murine model of disseminated candidiasis," *Microbes and Infection*, vol. 6, no. 6, pp. 542-548, 2004.

[88] C. Murciano, E. Villamon, D. Gozalbo, P. Roig, J. E. O'Connor, and M. L. Gil, "Toll-like receptor 4 defective mice carrying point or null mutations do not show increased susceptibility to *Candida albicans* in a model of hematogenously disseminated infection," *Medical Mycology*, vol. 44, no. 2, pp. 149-157, 2006.

[89] C. Bourgeois, O. Majer, I. E. Frohner et al., "Conventional dendritic cells mount a type I IFN response against Candida spp. requiring novel phagosomal TLR7-mediated IFN-$\beta$ signaling," *Journal of Immunology*, vol. 186, no. 5, pp. 3104-3112, 2011.

[90] A. Miyazato, K. Nakamura, N. Yamamoto et al., "Toll-like receptor 9-dependent activation of myeloid dendritic cells by deoxynucleic acids from *Candida albicans*," *Infection and Immunity*, vol. 77, no. 7, pp. 3056-3064, 2009.

[91] F. L. van de Veerdonk, M. G. Netea, T. J. Jansen et al., "Redundant role of TLR9 for anti-Candida host defense," *Immunobiology*, vol. 213, no. 8, pp. 613-620, 2008.

[92] C. Biondo, G. Signorino, A. Costa et al., "Recognition of yeast nucleic acids triggers a host-protective type I interferon response," *European Journal of Immunology*, vol. 41, no. 7, pp. 1969-1979, 2011.

[93] J. Magarian Blander and R. Medzhitov, "Toll-dependent selection of microbial antigens for presentation by dendritic cells," *Nature*, vol. 440, no. 7085, pp. 808-812, 2006.

[94] T. S. Plantinga, M.D. Johnson,W. K. Scott et al., "Human genetic susceptibility to Candida infections," *MedicalMycology*, vol. 50, no. 8, pp. 785-794, 2012.

[95] T. S. Plantinga,M.D. Johnson,W.K. Scott et al., "Toll-like receptor 1 polymorphisms increase susceptibility to candidemia," *Journal of Infectious Diseases*, vol. 205, no. 6, pp. 934-943, 2012.

[96] A. Nahum, H. Dadi, A. Bates, and C. M. Roifman, "The L412F variant of Toll-like receptor 3 (TLR3) is associated with cutaneous candidiasis, increased

susceptibility to cytomegalovirus, and autoimmunity," *Journal of Allergy and Clinical Immunology*, vol. 127, no. 2, pp. 528-531, 2011.

[97] A. Nahum, H. Dadi, A. Bates, and C. M. Roifman, "The biological significance of TLR3 variant, L412F, in conferring susceptibility to cutaneous candidiasis, CMV and autoimmunity," *Autoimmunity Reviews*, vol. 11, no. 5, pp. 341-347, 2012.

[98] C. A. A. Van der Graaf, M. G. Netea, S. A. Morr'e et al., "Tolllike receptor 4 Asp299Gly/Thr399Ile polymorphisms are a risk factor for Candida bloodstream infection," *European Cytokine Network*, vol. 17, no. 1, pp. 29-34, 2006.

[99] A. N. Zelensky and J. E. Gready, "The C-type lectin-like domain superfamily," *FEBS Journal*, vol. 272, no. 24, pp. 6179-6217, 2005.

[100] S. E. Hardison and G. D. Brown, "C-type lectin receptors orchestrate antifungal immunity," *Nature Immunology*, vol. 13, no. 9, pp. 817-822, 2012.

[101] J. A. Willment and G. D. Brown, "C-type lectin receptors in antifungal immunity," *Trends in Microbiology*, vol. 16, no. 1, pp. 27-32, 2008.

[102] G. D. Brown, "Dectin-1 : a signalling non-TLR patternrecognition receptor," *Nature Reviews Immunology*, vol. 6, no. 1, pp. 33-43, 2006.

[103] S. I.Gringhuis, J. denDunnen,M. Litjens et al., "Dectin-1 directs T helper cell differentiation by controlling noncanonical NF-κB activation through Raf-1 and Syk," *Nature Immunology*, vol. 10, no. 2, pp. 203-213, 2009.

[104] D. M. Reid, N. A. Gow, and G. D. Brown, "Pattern recognition: recent insights fromDectin-1," *Current Opinion in Immunology*, vol. 21, no. 1, pp. 30-37, 2009.

[105] J. Herre, J. A. Willment, S. Gordon, and G. D. Brown, "The role of dectin-1 in antifungal immunity," *Critical Reviews in Immunology*, vol. 24, no. 3, pp. 193-203, 2004.

[106] H. S. Goodridge, C. N. Reyes, C. A. Becker et al., "Activation of the innate immune receptor Dectin-1 upon formation of a 'Phagocytic synapse'," *Nature*, vol. 472, no. 7344, pp. 471-475, 2011.

[107] G. D. Brown, J. Herre, D. L. Williams, J. A. Willment, A. S. J. Marshall, and S. Gordon, "Dectin-1 mediates the biological effects of β-glucans," *Journal of ExperimentalMedicine*, vol. 197, no. 9, pp. 1119-1124, 2003.

[108] S. Dillon, S. Agrawal, K. Banerjee et al., "Yeast zymosan, a stimulus for TLR2 and dectin-1, induces regulatory antigenpresenting cells and immunological tolerance," *Journal of Clinical Investigation*, vol. 116, no. 4, pp. 916-928, 2006.

[109] B. N. Gantner, R. M. Simmons, S. J. Canavera, S. Akira, and D. M. Underhill, "Collaborative induction of inflammatory responses by dectin-1 and toll-like receptor 2," *Journal of Experimental Medicine*, vol. 197, no. 9, pp. 1107-1117, 2003.

[110] C. A. Rappleye, L. G. Eissenberg, and W. E. Goldman, "Histoplasma capsulatumα-(1,3)-glucan blocks innateimmune recognition by the β-glucan receptor," *Proceedings of the National Academy of Sciences of the United States of America*, vol. 104, no. 4, pp. 1366-1370, 2007.

[111] D. W. Lowman, R. R. Greene, D. W. Bearden et al., "Novel structural features in *Candida albicans* hyphal

glucan provide a basis for differential innate immune recognition of hyphae versus yeast," *The Journal of Biological Chemistry*, vol. 289, pp. 3432-3443, 2014.

[112] P. J. Rice, E. L. Adams, T. Ozment-Skelton et al., "Oral delivery and gastrointestinal absorption of soluble glucans stimulate increased resistance to infectious challenge," *Journal of Pharmacology and ExperimentalTherapeutics*, vol. 314, no. 3, pp. 1079-1086, 2005.

[113] S. E. Evans, P. Y. Hahn, F. McCann, T. J. Kottom, Z. V. Pavlovi'c, and A. H. Limper, "Pneumocystis cell wall $\beta$-glucans stimulate alveolar epithelial cell chemokine generation through nuclear factor-$\kappa\beta$-dependent mechanisms," *The American Journal of Respiratory Cell and Molecular Biology*, vol. 32, no. 6, pp. 490-497, 2005.

[114] H. Lee, J. Yuk, D. Shin, and E. Jo, "Dectin-1 is inducible and plays an essential role for mycobacteria-induced innate immune responses in airway epithelial cells," *Journal of Clinical Immunology*, vol. 29, no. 6, pp. 795-805, 2009.

[115] D. L.Moyes, M. Runglall, C.Murciano et al., "A biphasic innate immune MAPK response discriminates between the yeast and hyphal forms of *Candida albicans* in epithelial cells," *Cell Host and Microbe*, vol. 8, no. 3, pp. 225-235, 2010.

[116] D. L. Moyes, C. Shen, C. Murciano et al., "Protection against epithelial damage during *Candida albicans* infection is mediated by PI3K/Akt and mammalian target of rapamycin signaling," *Journal of InfectiousDiseases*, vol. 209, no. 11,pp. 1816-1826, 2014.

[117] J. Wagener, G. Weindl, P. W. J. de Groot et al., "Glycosylation of *Candida albicans* cell wall proteins is critical for induction of innateimmune responses and apoptosis of epithelial cells," *PLoS ONE*, vol. 7, no. 11,Article ID e50518, 2012.

[118] S. Saijo, N. Fujikado, T. Furuta et al., "Dectin-1 is required for host defense against Pneumocystis carinii but not against *Candida albicans*," *Nature Immunology*, vol. 8, no. 1, pp. 39-46, 2007.

[119] P. R. Taylor, S. V. Tsoni, J. A. Willment et al., "Dectin-1 is required for $\beta$-glucan recognition and control of fungal infection," *Nature Immunology*, vol. 8, no. 1, pp. 31-38, 2007.

[120] O.Gross, A.Gewies, K. Finger et al., "Card9 controls a non-TLR signalling pathway for innate anti-fungal immunity," *Nature*, vol. 442, no. 7103, pp. 651-656, 2006.

[121] A. Puel, S. Cypowyj, J. Bustamante et al., "Chronic mucocutaneous candidiasis in humans with inborn errors of interleukin-17 immunity," *Science*, vol. 332, no. 6025, pp. 65-68, 2011.

[122] D. C. Rosentul, T. S. Plantinga, M. Oosting et al., "Genetic variation in the dectin-1/CARD9 recognition pathway and susceptibility to candidemia," *Journal of Infectious Diseases*, vol. 204, no. 7, pp. 1138-1145, 2011.

[123] I. D. Iliev, V. A. Funari, K. D. Taylor et al., "Interactions between commensal fungi and the C-type lectin receptor dectin-1 influence colitis," *Science*, vol. 336, no. 6086, pp. 1314-1317, 2012.

[124] S. Vautier, R. A. Drummond, P. Redelinghuys, G. I. Murray, D. M. MacCallum, and G. D. Brown, "Dectin-1 is not required for controlling *Candida albicans* colonization of the gastrointestinal tract," *Infection and Immunity*, vol. 80,no. 12, pp. 4216-4222, 2012.

[125] B. Ferwerda, G. Ferwerda, T. S. Plantinga et al., "Human dectin-1 deficiency and mucocutaneous fungal infections," *The New England Journal ofMedicine*, vol. 361, no. 18, pp. 1760-1767, 2009.

[126] T. S. Plantinga, W. J. F. M. Van Der Velden, B. Ferwerda et al., "Early stop polymorphism in human DECTIN-1 is associated with increased candida colonization in hematopoietic stem cell transplant recipients," *Clinical Infectious Diseases*, vol. 49, no. 5, pp. 724-732, 2009.

[127] T. S. Plantinga, O. J. M. Hamza, J. A. Willment et al., "Genetic variation of innate immune genes in HIV-infected African patients with or without oropharyngeal candidiasis," *Journal of Acquired ImmuneDeficiency Syndromes*, vol. 55,no. 1, pp. 87-94, 2010.

[128] L.M. Grahamand G. D. Brown, "TheDectin-2 family of C-type lectins in immunity and homeostasis," *Cytokine*, vol. 48, no. 1-2, pp. 148-155, 2009.

[129] E. P. McGreal, M. Rosas, G. D. Brown et al., "The carbohydrate recognition domain of Dectin-2 is a C-type lectin with specificity for high mannose," *Glycobiology*, vol. 16, no. 5, pp. 422-430, 2006.

[130] S. Saijo, S. Ikeda, K. Yamabe et al., "Dectin-2 recognition of $\alpha$-mannans and induction of Th17 cell differentiation is essential for host defense against *Candida albicans*," *Immunity*, vol. 32, no. 5, pp. 681-691, 2010.

[131] M. J. Robinson, F. Osorio, M. Rosas et al., "Dectin-2 is a Syk coupled pattern recognition receptor crucial forTh17 responses to fungal infection," *Journal of Experimental Medicine*, vol. 206, no. 9, pp. 2037-2051, 2009.

[132] N. Hirata, K. Ishibashi, W. Sato et al., "Beta-mannosyl linkages inhibit CAWS arteritis by negatively regulating dectin-2-dependent signaling in spleen and dendritic cells," *Immunopharmacology and Immunotoxicology*, vol. 35, pp. 594-604, 2013.

[133] B. Kerscher, J. A. Willment, and G. D. Brown, "The Dectin-2 family of C-type lectin-like receptors: an update," *International Immunology*, vol. 25, no. 5, pp. 271-277, 2013.

[134] D. C. Ifrim, J. M. Bain,D. M. Reid et al., "The role of Dectin-2 for host defense against systemic infection with *Candida glabrata*," *Infection and Immunity*, vol. 82, no. 3, pp. 1064-1073, 2014.

[135] L. Zhu, X. Zhao, C. Jiang et al., "C-type lectin receptors dectin-3 and dectin-2 form a heterodimeric pattern-recognition receptor for host defense against fungal infection," *Immunity*, vol. 39, no. 2, pp. 324-334, 2013.

[136] A. Cambi, M. G. Netea, H. M. Mora-Montes et al., "Dendritic cell interaction with *Candida albicans* critically depends on Nlinked Mannan," *The Journal of Biological Chemistry*, vol. 283, no. 29, pp. 20590-20599, 2008.

[137] J. S. Lam, H. Huang, and S. M. Levitz, "Effect of differential N-linked and O-linked mannosylation on recognition of fungal antigens by dendritic cells," *PLoS ONE*, vol. 2, no. 10, Article ID e1009, 2007.

[138] P. R. Taylor, S. Gordon, and L. Martinez-Pomares, "The mannose receptor: linking homeostasis and immunity through sugar recognition," *Trends in Immunology*, vol. 26, no. 2, pp. 104-110, 2005.

[139] U. Gazi,M. Rosas, S. Singh et al., "Fungal recognition enhances mannose receptor shedding through dectin-1 engagement," *Journal of Biological*

*Chemistry*, vol. 286, no. 10, pp. 7822-7829, 2011.

[140] F. L. van de Veerdonk, R. J. Marijnissen, B. J. Kullberg et al., "The macrophage mannose receptor induces IL-17 in response to *Candida albicans*," *Cell Host and Microbe*, vol. 5, no. 4, pp. 329-340, 2009.

[141] S. E.M. Heinsbroek, P. R. Taylor, F. O.Martinez, L. Martinez- Pomares, G. D. Brown, and S. Gordon, "Stage-specific sampling by pattern recognition receptors during *Candida albicans* phagocytosis," *PLoS Pathogens*, vol. 4, no. 11, Article ID e1000218, 2008.

[142] J. Zhang, S. D. Tachado, N. Patel et al., "Negative regulatory role of mannose receptors on human alveolar macrophage proinflammatory cytokine release in vitro," *Journal of Leukocyte Biology*, vol. 78, no. 3, pp. 665-674, 2005.

[143] S. J. Lee,N. Zheng,M. Clavijo, andM.C.Nussenzweig, "Normal host defense during systemic candidiasis in mannose receptordeficient mice," *Infection and Immunity*, vol. 71, no. 1, pp. 437-445, 2003.

[144] J. M. Dan, R. M. Kelly, C. K. Lee, and S. M. Levitz, "Role of the mannose receptor in a murine model of *Cryptococcus neoformans* infection," *Infection and Immunity*, vol. 76, no. 6, pp. 2362-2367, 2008.

[145] F. Martinon, A. Mayor, and J. Tschopp, "The inflammasomes: guardians of the body," *Annual Review of Immunology*, vol. 27, pp. 229-265, 2009.

[146] H. Kumar, T. Kawai, and S. Akira, "Pathogen recognition by the innate immune system," *International Reviews of Immunology*, vol. 30, no. 1, pp. 16-34, 2011.

[147] C. Bryant and K. A. Fitzgerald, "Molecular mechanisms involved in inflammasome activation," *Trends in Cell Biology*, vol. 19, no. 9, pp. 455-464, 2009.

[148] C. A. A. Van Der Graaf, M. G. Netea, B. Franke, S. E. Girardin, J.W. M. VanDerMeer, and B. J. Kullberg, "Nucleotide oligomerization domain 2 (Nod2) is not involved in the pattern recognition of*Candida albicans*," *Clinical andVaccine Immunology*, vol. 13, no. 3, pp. 423-425, 2006.

[149] S. Joly, N. Ma, J. J. Sadler, D. R. Soll, S. L. Cassel, and F. S. Sutterwala, "Cutting edge: *Candida albicans* hyphae formation triggers activation of the Nlrp3 inflammasome," *Journal of Immunology*, vol. 183, no. 6, pp. 3578-3581, 2009.

[150] J. Tomalka, S. Ganesan, E. Azodi et al., "A novel role for the NLRC4 inflammasome in mucosal defenses against the fungal pathogen *Candida albicans*," *PLoS Pathogens*, vol. 7, no. 12, Article ID e1002379, 2011.

[151] J. A. Kummer, R. Broekhuizen,H. Everett et al., "Inflammasome componentsNALP 1 and 3 showdistinct but separate expression profiles in human tissues suggesting a site-specific role in the inflammatory response," *Journal of Histochemistry and Cytochemistry*, vol. 55, no. 5, pp. 443-452, 2007.

[152] Y.Mostefaoui, I.Claveau, andM. Rouabhia, "In vitro analyses of tissue structure and interleukin-1$\beta$ expression and production by human oral mucosa in response to *Candida albicans* infections," *Cytokine*, vol. 25, no. 4, pp. 162-171, 2004.

[153] M. Rouabhia, G. Ross, N. Pag'e, and J. Chakir, "Interleukin-18 and gamma interferon production by oral epithelial cells in response to exposure to *Candida albicans* or lipopolysaccharide stimulation," *Infection and Immunity*, vol. 70, no. 12, pp. 7073-7080, 2002.

[154] G.Weindl, J. R. Naglik, S. Kaesler et al., "Human epithelial cells establish direct antifungal defense through TLR4-mediated signaling," *The Journal of Clinical Investigation*, vol. 117, no. 12, pp. 3664-3672, 2007.

[155] F. Tardif, J. Goulet, A. Zakrazewski, P. Chauvin, and M. Rouabhia, "Involvement of interleukin-18 in the inflammatory response against oropharyngeal candidiasis," *Medical Science Monitor*, vol. 10, no. 8, pp. BR239–BR249, 2004.

[156] O. Gross, H. Poeck, M. Bscheider et al., "Syk kinase signalling couples to the Nlrp3 inflammasome for anti-fungal host defence," *Nature*, vol. 459, no. 7245, pp. 433-436, 2009.

[157] F. L. van de Veerdonk, B. J. Kullberg, J. W. van der Meer, N. A. Gow, and M. G. Netea, "Host-microbe interactions: innate pattern recognition of fungal pathogens," *Current Opinion in Microbiology*, vol. 11, no. 4, pp. 305-312, 2008.

[158] A. G. Vonk,M.G.Netea, J.H.VanKrieken, Y. Iwakura, J.W. M. VanDerMeer, and B. J.Kullberg, "Endogenous interleukin (IL)-1$\alpha$ and IL-1$\beta$ are crucial for host defense against disseminated candidiasis," *The Journal of Infectious Diseases*, vol. 193, no. 10, pp. 1419-1426, 2006.

[159] A. G. Hise, J. Tomalka, S. Ganesan et al., "An essential role for the NLRP3 inflammasome in host defense against the human fungal pathogen *Candida albicans*," *Cell Host and Microbe*, vol. 5, no. 5, pp. 487-497, 2009.

[160] M. H. Zaki,K.L.Boyd, P. Vogel,M. B. Kastan,M.Lamkanfi, and T. Kanneganti, "The NLRP3 Inflammasome Protects against Loss of Epithelial Integrity and Mortality during Experimental Colitis," *Immunity*, vol. 32, no. 3, pp. 379-391, 2010.

[161] L. M. Rehaume, T. Jouault, andM. Chamaillard, "Lessons from the inflammasome: a molecular sentry linking Candida and Crohn's disease," *Trends in Immunology*, vol. 31, no. 5, pp. 171-175, 2010.

[162] A. Lev-Sagie, D. Prus, I. M. Linhares, Y. Lavy, W. J. Ledger, and S. S. Witkin, "Polymorphism in a gene coding for the inflammasome component NALP3 and recurrent vulvovaginal candidiasis in women with vulvar vestibulitis syndrome," *The American Journal of Obstetrics and Gynecology*, vol. 200, no. 3, pp. 303. e1-303.e6, 2009.

[163] S. Yamasaki, M. Matsumoto, O. Takeuchi et al., "C-type lectin Mincle is an activating receptor for pathogenic fungus, Malassezia," *Proceedings of the National Academy of Sciences of the United States of America*, vol. 106, no. 6, pp. 1897-1902, 2009.

[164] A. Bugarcic, K. Hitchens, A. G. Beckhouse, C. A. Wells, R. B. Ashman, and H. Blanchard, "Human and mouse macrophageinducible C-type lectin (Mincle) bind *Candida albicans*," *Glycobiology*, vol. 18, no. 9, pp. 679-685, 2008.

[165] D. Vijayan, K. J. Radford, A. G. Beckhouse, R. B. Ashman, and C. A.Wells, "Mincle polarizes humanmonocyte and neutrophil responses to *Candida albicans*," *Immunology and Cell Biology*, vol. 90, no. 9, pp. 889-895, 2012.

[166] *M. Da* Gl'oria Sousa, D. M. Reid, E. Schweighoffer et al., "Restoration of pattern recognition receptor costimulation to treat chromoblastomycosis, a chronic fungal infection of the skin," *Cell Host and Microbe*, vol. 9, no. 5, pp. 436-443, 2011.

[167] W. Lee, J. Kang, J. Yan et al., "Neutrophils promote mycobacterial trehalose dimycolate-induced lung

inflammation via the mincle pathway," *PLoS Pathogens*, vol. 8, no. 4, Article ID e1002614, 2012.

[168] P. F. Zipfel, "Complement and immune defense: From innate immunity to human diseases," *Immunology Letters*, vol. 126, no. 1-2, pp. 1-7, 2009.

[169] P. F. Zipfel and C. Skerka, "Complement, Candida, and cytokines: the role of C5a in host response to fungi," *European Journal of Immunology*, vol. 42, no. 4, pp. 822-825, 2012.

[170] S. Luo, C. Skerka, O. Kurzai, and P. F. Zipfel, "Complement and innate immune evasion strategies of the human pathogenic fungus *Candida albicans*,"*Molecular Immunology*, vol. 56, no. 3, pp. 161-169, 2013.

[171] C. Speth, G. Rambach, R. W¨urzner, and C. Lass-Fl¨orl, "Complement and fungal pathogens: an update," *Mycoses*, vol. 51, no. 6, pp. 477-496, 2008.

[172] N. Brouwer, K. M. Dolman, M. van Houdt, M. Sta, D. Roos, and T. W. Kuijpers, "Mannose-binding lectin (MBL) facilitates opsonophagocytosis of yeasts but not of bacteria despite MBL binding," *Journal of Immunology*, vol. 180, no. 6, pp. 4124-4132, 2008.

[173] O. Neth, D. L. Jack, A. W. Dodds, H. Holzel, N. J. Klein, and M. W. Turner, "Mannose-binding lectin binds to a range of clinically relevant microorganisms and promotes complement deposition," *Infection and Immunity*, vol. 68, no. 2, pp. 688-693, 2000.

[174] C. B. Forsyth and H. L. Mathews, "Lymphocytes utilize CD11b/CD18 for adhesion to *Candida albicans*," *Cellular Immunology*, vol. 170, no. 1, pp. 91-100, 1996.

[175] T. Meri, A. Hartmann, D. Lenk et al., "The yeast *Candida albicans* binds complement regulators factor H and FHL-1," *Infection and Immunity*, vol. 70, no. 9, pp. 5185-5192, 2002.

[176] T. Meri, A.M. Blom, A. Hartmann, D. Lenk, S. Meri, and P. F. Zipfel, "The hyphal and yeast forms of *Candida albicans* bind the complement regulator C4b-binding protein," *Infection and Immunity*, vol. 72, no. 11, pp. 6633-6641, 2004.

[177] V. Agarwal, T. M. Asmat, S. Luo, I. Jensch, P. F. Zipfel, and S. Hammerschmidt, "Complement regulator factor H mediates a two-step uptake of Streptococcus pneumoniae by human cells," *The Journal of Biological Chemistry*, vol. 285, no. 30, pp. 23486-23495, 2010.

[178] S. Poltermann, A. Kunert, M. Von Der Heide, R. Eck, A. Hartmann, and P. F. Zipfel, "Gpm1p is a factor H-, FHL-1-, and plasminogen-binding surface protein of *Candida albicans*," *Journal of Biological Chemistry*, vol. 282, no. 52, pp. 37537-37544, 2007.

[179] S. Luo, S. Poltermann, A. Kunert, S. Rupp, and P. F. Zipfel, "Immune evasion of the human pathogenic yeast *Candida albicans*: Pra1 is a Factor H, FHL-1 and plasminogen binding surface protein,"*Molecular Immunology*, vol. 47,no. 2-3, pp. 541-550, 2009.

[180] S.Luo,A.M.Blom, S. Ruppet al., "ThepH-regulatedantigen1of *Candida albicans* binds the human complement inhibitor C4bbinding protein and mediates fungal complement evasion," *Journal of Biological Chemistry*, vol. 286, no. 10, pp. 8021-8029, 2011.

[181] S. Luo, A. Hartmann, H. Dahse, C. Skerka, and P. F. Zipfel, "Secreted pH-regulated antigen 1 of *Candida albicans* blocks activation and conversion of complement C3," *Journal of Immunology*, vol. 185, no. 4, pp. 2164-2173, 2010.

[182] R. B. Ashman, J. M. Papadimitriou, A. Fulurija et al., "Role of complement C5 and T lymphocytes in pathogenesis of disseminated and mucosal candidiasis in susceptible DBA/2 mice," *Microbial Pathogenesis*, vol. 34, no. 2, pp. 103-113, 2003.

[183] R. B. Ashman, E. M. Bolitho, and J.M. Papadimitriou, "Patterns of resistance to *Candida albicans* in inbred mouse strains," *Immunology and Cell Biology*, vol. 71, no. 3, pp. 221-225, 1993.

[184] R. B. Ashman, "Genetic determination of susceptibility and resistance in the pathogenesis of *Candida albicans* infection," *FEMS Immunology and Medical Microbiology*, vol. 19, no. 3, pp. 183-189, 1997.

[185] I. Radovanovic, A. Mullick, and P. Gros, "Genetic control of susceptibility to infection with *Candida albicans* in mice," *PLoS ONE*, vol. 6, no. 4,Article ID e18957, 2011.

[186] A. Mullick, *M. Elias*, S. Picard et al., "Dysregulated inflammatory response to *Candida albicans* in a C5-deficient mouse strain," *Infection and Immunity*, vol. 72, no. 10, pp. 5868-5876, 2004.

[187] A. Mullick, Z. Leon, G. Min-Oo et al., "Cardiac failure in C5- deficient A/J mice after *Candida albicans* infection," *Infection and Immunity*, vol. 74, no. 8, pp. 4439-4451, 2006.

[188] S.V. Tsoni, A.M. Kerrigan,M. J.Marakalala et al., "Complement C3 plays an essential role in the control of opportunistic fungal infections," *Infection and Immunity*, vol. 77, no. 9, pp. 3679-3685, 2009.

[189] C. B. Forsyth and H. L. Mathews, "Lymphocyte adhesion to *Candida albicans*," *Infection and Immunity*, vol. 70, no. 2, pp. 517-527, 2002.

[190] W. K. Ip and Y. L. Lau, "Role of mannose-binding lectin in the innate defense against *Candida albicans*: enhancement of complement activation, but lack of opsonic function, in phagocytosis by human dendritic cells," *Journal of Infectious Diseases*, vol. 190, no. 3, pp. 632-640, 2004.

[191] M. C. Ghezzi, G. Raponi, S. Angeletti, and C.Mancini, "Serummediated enhancement of TNF-$\alpha$ release by human monocytes stimulated with the yeast form of *Candida albicans*," *Journal of Infectious Diseases*, vol. 178, no. 6, pp. 1743-1749, 1998.

[192] A. Sonesson, L. Ringstad, E. Andersson Nordahl, M.Malmsten, M. M¨orgelin, and A. Schmidtchen, "Antifungal activity of C3a and C3a-derived peptides against Candida," *Biochimica et Biophysica Acta—Biomembranes*, vol. 1768, no. 2, pp. 346-353, 2007.

[193] P. Miram'on, L. Kasper, and B. Hube, "Thriving within the host: candida spp. interactions with phagocytic cells," *Medical Microbiology and Immunology*, vol. 202, no. 3, pp. 183-195, 2013.

[194] E. Svobodov'a, P. Staib, J. Losse, F. Hennicke, D. Barz, and M. J'ozsi, "Differential interaction of the two related fungal species *Candida albicans* and Candida dubliniensis with human neutrophils," *Journal of Immunology*, vol. 189, no. 5, pp. 2502-2511, 2012.

[195] J. R. Linden, M. A. MacCani, S. S. Laforce-Nesbitt, and J. M Bliss, "High efficiency opsonin-independent phagocytosis of *Candida parapsilosis* by human neutrophils," *MedicalMycology*, vol. 48, no. 2, pp. 355-364, 2010.

[196] W. L. Lee, R. E. Harrison, and S. Grinstein, "Phagocytosis by neutrophils," *Microbes and Infection*, vol. 5, no. 14, pp. 1299-1306, 2003.

[197] B. Amulic, C. Cazalet, G. L. Hayes, K. D. Metzler, and A. Zychlinsky,

"Neutrophil function: From mechanisms to disease," *Annual Review of Immunology*, vol. 30, pp. 459-489, 2012.

[198] B. H. Segal, M. J. Grimm, A. N. H. Khan, W. Han, and T. S. Blackwell, "Regulation of innateimmunity byNADPHoxidase," *FreeRadical Biology andMedicine*, vol. 53,no. 1, pp. 72-80, 2012.

[199] F. C. Fang, "Antimicrobial reactive oxygen and nitrogen species: concepts and controversies," *Nature Reviews Microbiology*, vol. 2, no. 10, pp. 820-832, 2004.

[200] C. C. Winterbourn, M. B. Hampton, J. H. Livesey, and A. J. Kettle, "Modeling the reactions of superoxide and myeloperoxidase in the neutrophil phagosome: implications for microbial killing," *Journal of Biological Chemistry*, vol. 281, no. 52, pp. 39860-39869, 2006.

[201] C. F. Urban, U. Reichard, V. Brinkmann, and A. Zychlinsky, "Neutrophil extracellular traps capture and kill *Candida albicans* and hyphal forms," *Cellular Microbiology*, vol. 8, no. 4, pp. 668-676, 2006.

[202] C. F. Urban, D. Ermert, M. Schmid et al., "Neutrophil extracellular traps contain calprotectin, a cytosolic protein complex involved in host defense against *Candida albicans*," *PLoS Pathogens*, vol. 5, no. 10, Article IDe1000639, 2009.

[203] K. Seider, A. Heyken, A. L¨uttich, P. Miram'on, and B. Hube, "Interaction of pathogenic yeasts with phagocytes: survival, persistence and escape," *Current Opinion in Microbiology*, vol. 13, no. 4, pp. 392-400, 2010.

[204] P. Perumal, S. Mekala, C. Nombela, W. L. Chaffin, and C. Gil, "Proteomic analysis of cytoplasmic and surface proteins from yeast cells, hyphae, and biofilms of *Candida albicans*," *Proteomics*, vol. 9, no. 8, pp. 2230-2252, 2009.

[205] F. O. Martinez, L. Helming, and S. Gordon, "Alternative activation of macrophages: an immunologic functional perspective," *Annual Review of Immunology*, vol. 27, pp. 451-483, 2009.

[206] S. Gordon and F. O. Martinez, "Alternative activation of macrophages: mechanism and functions," *Immunity*, vol. 32,no. 5, pp. 593-604, 2010.

[207] C. Jim'enez-L'opez and M. C. Lorenz, "Fungal immune evasion in a model host-pathogen interaction: *Candida albicans* versus macrophages," *PLoS Pathogens*, vol. 9, Article IDe1003741, 2013.

[208] O. V. Vieira, R. J. Botelho, and S. Grinstein, "Phagosome maturation: aging gracefully," *Biochemical Journal*, vol. 366, no. 3, pp. 689-704, 2002.

[209] D. M. Mosser and J. P. Edwards, "Exploring the full spectrum of macrophage activation," *Nature Reviews Immunology*, vol. 8, no. 12, pp. 958-969, 2008.

[210] M. Stein, S. Keshav, N. Harris, and S. Gordon, "Interleukin 4 potently enhances murine macrophage mannose receptor activity: a marker of alternative immunologic macrophage activation," *Journal of ExperimentalMedicine*, vol. 176, no. 1, pp. 287-292, 1992.

[211] M.Hesse,M.Modolell, A. C. La Flamme et al., "Differential regulation of nitric oxide synthase-2 and arginase-1 by type 1/type 2 cytokines in vivo: granulomatous pathology is shaped by the pattern of L-arginine metabolism," *Journal of Immunology*, vol.167, no. 11, pp. 6533-6544, 2001.

[212] A. Tavanti, D. Campa, A. Bertozzi et al., "*Candida albicans* isolates with different genomic backgrounds display a

differential response to macrophage infection," *Microbes and Infection*, vol. 8, no. 3, pp. 791-800, 2006.

[213] I. E. Frohner, C. Bourgeois, K. Yatsyk,O.Majer, andK. Kuchler, "*Candida albicans* cell surface superoxide dismutases degrade host-derived reactive oxygen species to escape innate immune surveillance,"*Mol ecularMicrobiology*, vol. 71,no. 1, pp. 240-252, 2009.

[214] B. D. Ullmann, H. Myers, W. Chiranand et al., "Inducible defense mechanism against nitric oxide in *Candida albicans*," *Eukaryotic Cell*, vol. 3, no. 3, pp. 715-723, 2004.

[215] E. Fern'andez-Arenas, V. Cabez'on, C. Bermejo et al., "Integrated proteomics and genomics strategies bring new insight into *Candida albicans* response upon macrophage interaction," *Molecular and Cellular Proteomics*, vol. 6, no. 3, pp. 460-478, 2007.

Section 2

# Mechanism of Molecular, Metabolic and Immune Response on *Candida albicans* Infection

Chapter 2

# Molecular Mechanisms of Resistance to Antifungals in *Candida albicans*

*Estela Ruiz-Baca, Rosa Isela Arredondo-Sánchez, Karina Corral-Pérez, Angélica López-Rodríguez, Iván Meneses-Morales, Víctor M. Ayala-García and Ana Lilia Martínez-Rocha*

## Abstract

Invasive Candidiasis (IC) presents a global mortality rate greater than 40%, occupying the fourth place worldwide as the most frequent opportunistic nosocomial disease. Although the genus *Candida* consists of around 200 species, only 20 are reported as etiological agents of IC, being *Candida albicans* the most frequent causal agent. Even when there is a broad range of antifungals drugs for *Candida* infections, azoles, polyenes, and echinocandins are considered among the most effective treatment. However, there is some incidence for antifungal resistance among some *Candida* strains, limiting treatment options. Several molecular mechanisms with antifungal agents have been reported for *C. albicans* where insertions, deletions, and point mutations in genes codifying target proteins are frequently related to the antifungal drug resistance. Furthermore, gene overexpression is also frequently associated to antifungal resistance as well as an increase in the activity of proteins that reduce oxidative damage. This chapter summarizes the main molecular mechanisms to *C. albicans* antifungal drug resistance, besides offering an overview of new antifungal agents and new antifungal targets to combat fungal infections.

**Keywords:** resistance mechanism, antifungal, azoles, polyenes, echinocandins

## 1. Introduction

*Candida albicans* is the most important opportunistic commensal yeast that asymptomatically colonizes the skin, oral cavity, gastrointestinal and genitourinary tracts in healthy people. However, it can cause superficial and invasive infections, especially in immunocompromised individuals [1–3]. Actually, invasive infections due to *Candida* species are considered among the main causes of morbidity and mortality in hospitalized patients. Although there are at least 15 *Candida* species related to human disease, more than 90% of the invasive diseases are related to *C. albicans*, *Candida glabrata*, *Candida tropicalis*, *Candida parapsilosis*, and *Candida krusei* [4–6]. *C. albicans* infections is considered the fourth most common



IntechOpen

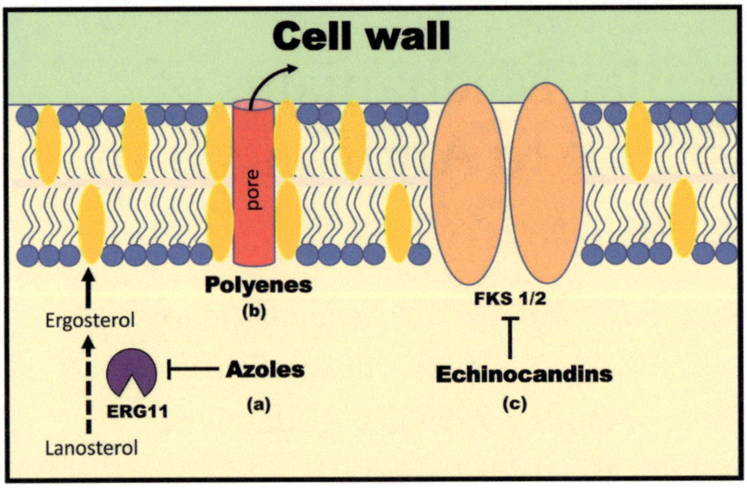


**Figure 1.**
*Mechanisms of action of main antifungals families in the fungal cell. (a) Azoles disrupt the ergosterol synthesis by inhibiting the enzyme 14-α-lanosterol demethylase (ERG11) involved in the transformation of lanosterol into ergosterol in the endoplasmic reticulum. (b) Polyenes disrupt the cell membrane by binding to ergosterol resulting in pore formation. (c) Echinocandins inhibit 1,3-β-d-glucan synthase (FKS ½) which causes disruption of the cell wall.*

opportunistic infection in hospitals. Invasive candidiasis (IC) is fatal in about 42% of the reported cases, despite the use of antifungal therapies [7, 8].

Nowadays, the most widely used antifungal drugs for IC include: A) azoles, for instance fluconazole (FLZ), itraconazole (ITC), voriconazole, posaconazole, isavuconazole; B) polyenes such as amphotericin B (AMB); C) echinocandins like caspofungin, micafungin, and anidulafungin [9–11].

These antifungal compounds act on different parts of the fungal cell (**Figure 1**). Azoles interrupt the ergosterol biosynthesis, the main component of the fungal membranes [10, 12, 13]. Polyenes such as AMB interact with ergosterol making pores in the cell membrane [10, 12–14]; while echinocandins act blocking the synthesis of β-d-glucan located in the fungal cell wall [13, 15]. The gradual risk increment for Candida infection and the greater use of antifungal agents has increased resistance towards *Candida* spp. Pharmacological failures in *Candida* spp. treatments have drawn attention to the problem of resistance to antifungals and their molecular mechanisms. *C. albicans* inherently is susceptible to azoles, polyenes, and echinocandins. Mono-resistance to azoles or echinocandins has been reported, as well as combined resistance to azoles and amphotericin, but resistance to multiple compounds that covers all three drug classes is a rare phenomenon and few cases have been reported in *C. albicans* [10, 12, 16].

The following chapter offers an overview of the main genetic mechanisms contributing to the antifungal resistance in *C. albicans*, besides giving an approach for alternative-compounds proposed against their infection.

## 2. Molecular mechanisms of antifungal resistance

### 2.1 Azoles

Fungi cell membrane is mainly integrated by ergosterol, a sterol contributing to several cellular functions, besides modulating membrane fluidity and the structure and function of membrane proteins. The azoles mechanism of action is to inhibit

14α-lanosterol demethylase, encoded by the *ERG11* gene, which converts lanosterol to ergosterol in the cell membrane (**Figure 1**). This enzyme contains an iron protoporphyrin unit in its active site. Azoles bind to iron causing the blockage of the ergosterol biosynthetic pathway [17–19]. The interruption of ergosterol synthesis allows the accumulation of 14α-methyl sterols, which alters the membrane's stability, permeability, and the action of the enzymes bound to it [20].

The evolution of antimicrobial agent's resistance is common, as there are many microbes able to develop strategies against drugs action. The incremented azoles resistance is mainly a result of their fungistatic rather than fungicidal nature [17–19]. The mechanisms of resistance to azole antifungal agents have been elucidated in *Candida* spp. species and can be classified mainly as: 1) changes in cell wall or in plasma membrane, leading to poor drug absorption; 2) alterations in the affinity of the target drug (i.e. *ERG11* gene), due to a site mutation or its overexpression; 3) drug efflux mediated by membrane transporter proteins belonging to the transporters of the ATP-binding cassette (ABC), namely CDR1 and CDR2 or the transporter of the major facilitator superfamily (MFS), CaMDR1; 4) biofilm formation [18–21]. Although the resistance described in *C. albicans* strains is usually a combination of the mechanisms mentioned above (**Figure 2**) [16].

### *2.1.1 Mutations of the ERG11 target enzyme*

Mutations in the *C. albicans ERG11* gene reduce the affinity for fluconazole and have a moderate effect on posaconazole [17–19]. Several point mutations have been identified in the *ERG11* gene. In resistant strains, there are more than 140 substitutions reported, most of them have a functional additive effect. Two of the most common alterations in *C. albicans* (R467K and G464S), are located near the heme-binding site [20]. Other substitutions related to resistance are A114S, Y132H, Y132F, K143R, Y257H, and K143Q, which contribute to a significant increased resistance (more than four times) to fluconazole and voriconazole [22].

Some clinical isolates share common mutations with environmental azole-resistant strains, suggesting that some azole-resistant clinical isolates could have their origin in the environment [23]. This resistance appears to be driven by the

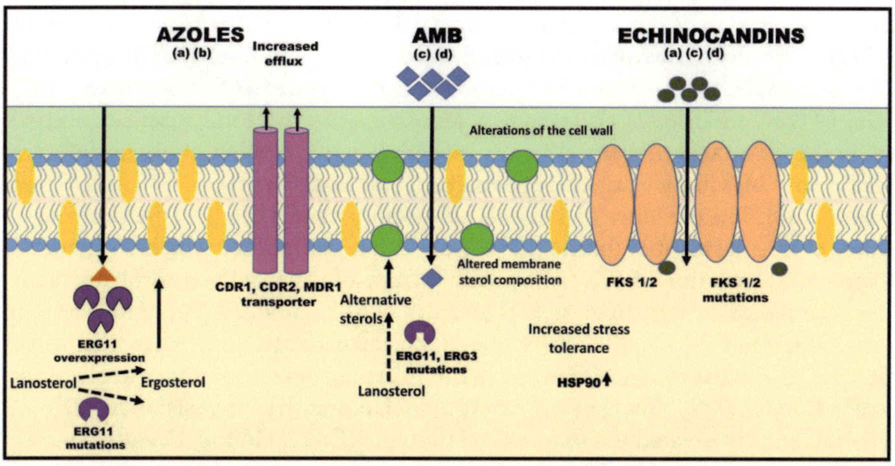


**Figure 2.**
*Schematic overview of the main mechanisms of drug resistance against azoles, AMB, and echinocandins adopted by* Candida albicans. *(a) Alteration of the enzyme target (azoles and echinocandins), (b) overexpression of drug efflux proteins (azoles), (c) Reduction of sterols in the plasma membrane (AMB), (d) increased stress tolerance and altered the fungal cell wall (echinocandins and AMB).*

agricultural use of azoles. In patients without azoles treatment, resistance has been identified derived from the environment. These cases involved a Cyp51A substitution at position 98 (from leucine to histidine), and a 34 base tandem repeat (TR) in the cyp51A promoter, leading to overexpression. Both changes are necessary to confer resistance. In particular, these resistant isolates can be crossed with susceptible strains, suggesting that resistance could be transferred through the sexual cycle. Strains with these alterations have emerged throughout Europe and beyond. Additionally, a new environmentally selected resistance mutation (TR46, Y121F, T289A) was reported among patients in the Netherlands [20].

### *2.1.2 Dysregulation of the target enzyme ERG11*

One way to decrease the drug effective concentration is the overexpression of *ERG11* [17]. This overexpression is common among azole-resistant *C. albicans* clinical isolates. This contributes directly to resistance, since an increase in the target requires more drug for inhibition, reducing susceptibility [19]. *ERG11* overexpression arises either from genetic dosing through gene duplication or from positive regulation of the gene by trans-acting factors [23]. Multiple mechanisms explain the constitutive overexpression of *ERG11* in azole-resistant clinical strains. First, amplification of the *ERG11* gene can occur by the formation of an isochromosome with two copies of the left arm of chromosome 5 [i (5 L)], in which *ERG11* resides, or by duplication of the entire chromosome. Second, the activation of mutations in the gene encoding the transcription factor Upc2 positively regulates most of the ergosterol biosynthesis genes [18, 20].

### *2.1.3 Alteration of the ergosterol biosynthesis pathway (point mutations in ERG genes)*

Brief exposures of two to three hours to azoles cause transient upregulation of the *ERG* gene family in *C. albicans.* These data suggest a common regulation of ergosterol biosynthetic pathway in the presence of inhibitors. Longer *in vitro* exposures to azoles (minimum 24 h) leads to constitutive up-regulation of the *ERG* genes decreasing drug susceptibility [23].

Modification of the metabolic pathway can be effective at different points, as example, alteration of the last steps of biosynthesis through the inactivation of the *ERG3* gene results in no toxic methylated sterols production, leading to azole cross-resistance. Furthermore, mutations in non-essential genes of this pathway (*ERG3, ERG6, ERG24,* and *ERG2*) also lead to a decrease, or even a total absence, of ergosterol in the plasma membrane [17]. Lanosterol demethylase inactivity or defectiveness due to azoles induce ergosterol depletion and toxic 14α-methyl-3,6-diol sterols accumulation. The presence of 14α-methyl sterols can modify the function and fluidity of the plasma membrane [21]. The additive mutation in the *ERG3* gene prevents the formation of this toxic product from 14α-methylfecosterol and leads to the accumulation of non-toxic sterols (Mishra, 2007; Shukla, 2016). Although this mechanism is not the most frequent one, it has been identified in several clinical isolates of *C. albicans* [23]. Mutations in *ERG3* are sufficient to induce azole resistance in *Candida* spp., but they are rarely associated with high resistance [20].

Four clinically isolated *C. albicans* erg3 mutants (CA12, CA488, CA490, and CA108) were reported as resistant to fluconazole, voriconazole, itraconazole, ketoconazole, and clotrimazole under CLSI test conditions. Importantly, CA12 and CA108 retained an azole-resistant phenotype even when tested in the presence of FK506, a multi-drug flux inhibitor. In contrast, CA488, CA490, along with three isolates (CA6, CA14, and CA177, in which ergosterol comprised more than 80% of the total sterol fraction and ergosta 7,22-dienol was undetectable) exhibited azole sensitive phenotypes in the inhibitor FK50 presence. CA108 mutant strain contains multiple amino acid substitutions in ERG3, but only a single conserved polymorphism (E266D) in sterol 14α-demethylase (ERG11). CA12 contains a substitution (W332R) in ERG3 and no residue changes in ERG11. Furthermore, CA488 and CA490 were found to harbour multiple residue changes in both ERG3 and ERG11 [24]. Furthermore, the residue 193 in ERG3 was found to play an important role in azole resistance [25].

#### *2.1.4 Efflux pumps*

A mechanism to decrease the azoles intracellular concentration is increasing their output. This class of resistance is mediated by the activity of transport systems such as the pleiotropic drug resistance (PDR) class of ATP-binding cassette transporters (ABC) and major facilitators superfamily (MFS) transporters [17]. These membrane proteins translocate compounds across cell membranes actively using different energy sources. ABC proteins are primary transporters that use ATP hydrolysis. MFS pumps are secondary transporters that use the motive force of the proton across the plasma membrane. Both types of transporters contain distinctive protein domains that confer substrate specificity: nucleotide-binding domains (NBD) in ABC pumps and transmembrane domains (TMD) in ABC and MFS pumps. Fungal PDR proteins appear to share common features on both sides of the two TMDs that separate the cytosolic from the outer cytosolic space [18, 26]. This probably reflects the fact that the cytosolic part is the motor that drives the transport of a variety of substrates through the lipid bilayer through the core of the protein into the outer cytosolic space or the outer layer of the lipid bilayer [26].

*C. albicans* contains 28 ABC proteins and 96 potential MFS transporters [18]. In this species, the main transporters, related to resistance, of the ABC proteins are CDR1 and CDR2 (resistance drugs to Candida 1 and 2) [21], while for MFS it is MDR1 (Multidrug Resistance 1). CDR1 and CDR2 overexpression improves drug output and reduces its accumulation in cells [23]. Positive regulation of MDR1 results in increased azole output [17]. Several cis-acting regulatory elements responsible for the regulation of the *CDR1* and *CDR2* genes have been identified. Promoter deletion studies have revealed five different regulatory elements in the CDR1 promoter, including one BEE (basal expression element), one DRE (drug-sensitive element), two SRE (steroid sensitive element), and one NRE (negative regulatory element). Internal deletions of the BEE and DRE motifs in the CDR1 promoter affect baseline CDR1 expression and drug-induced expression, respectively. SRE1 and SRE2 are involved in steroid hormone responses: SRE1 responds only to progesterone and SRE2 to progesterone and β-estradiol. Finally, the deletion of the NRE motif leads to an increase in the baseline expression of CDR1. In contrast to CDR1, the CDR2 promoter contains only one DRE motif. Among these diverse cis-acting elements, DRE is the only element involved in constitutive high expression and transient up-regulation of CDR1 and CDR2. In *C. albicans*, CDR1 is the main contributor to azole resistance among ABC transporters [23, 26].

In *C. albicans* a gene encoding a CaNdt80p protein similar to the *Saccharomyces cerevisiae* meiosis-specific transcription factor Ndt80p has been identified. Alteration of CaNdt80 affects the basal expression of CDR1 and reduces its ability for up-regulation in the presence of miconazole. More recently, Ndt80p was involved in the global effect of azole resistance through its regulon, including several genes implicated in ergosterol metabolism [23]. Additionally, MDR1 is the only MFS transporter involved in the azole resistance of clinical isolates. MDR1 usually does not express detectable levels in fluconazole-susceptible isolates but is constitutively up-regulated in some fluconazole-resistant strains. A region called BRE (benomyl response element) or MDRE (*MDR1* drug resistance element), respectively, was identified. This region is responsible for the constitutively high expression of *MDR1* in fluconazole-resistant isolates. Hyperactive alleles confer a constitutive overexpression of *MDR1* and therefore, resistance to fluconazole [23]. *MDR1* expression in *C. albicans* cells is enhanced by benomyl, methotrexate, and several other unrelated drugs, and found to be more pronounced in some of the azole-resistant clinical isolates [21].

The up-regulation of ABC and MFS transporters is mediated by specific regulations in resistant fungal pathogens. Point mutations defined as gain-of-function (GOF) mutations in these regulators confer an inherently high level of expression of the transporters in drug-resistant strains. GOF mutations in the transcription factor Upc2p led to increased resistance to fluconazole in *C. albicans* [17]. GOF mutations in the transcription factors TAC1 and MRR1 lead to upregulation of the CDR1/CDR2 and MDR1 drug efflux pumps, respectively [16, 18]. An important question related to strategies to overcome efflux-mediated antifungal resistance is the relative contribution of each efflux pump protein to clinically significant antifungal resistance in *C. albicans*. It is now clear that the CDR1, CDR2, and MDR1 transporters are the main efflux pumps that mediate resistance of *C. albicans* to azole drugs. However, MDR1 is relatively specific for fluconazole, while many azole drugs can act as substrates for CDR1 and CDR2. Interestingly, several fluconazole-resistant *C. albicans* isolates overexpress only CDR1 and CDR2, but not MDR1, while other strains overexpress only MDR1, reflecting the existence of at least two different transcriptional pathways that are responsible for the upregulation of these genes in azoles [26].

### 2.2 Polyenes

The potent fungicidal activity of polyenes derives from their ability to selectively bind sterol at the fungal cell membrane (**Figure 1**). Four models have been proposed as the mode of action for polyenes: 1) the pore formation model, 2) the surface adsorption model, 3) the sterol sponge model, and 4) the oxidative damage model [14]. The pore formation model is the most studied mechanism, where polyenes are directly intercalated with the ergosterol membrane forming ion channels that permeabilize and kill yeast cells [14, 27]. Additionally, indirect mechanisms of fungal cells damage have been identified due to the effect of polyene compounds, such as those mediated by reactive oxygen species (ROS) and by the secretion of interleukin-1β (IL-1β) by host cells [28, 29].

The polyene AMB is a broad-spectrum drug and is one of the main antifungals used for ICs [10, 14]. AMB is heptane isolated from *Streptomyces nodosus* producing high toxicity. Hence, a liposomal AMB (Ambisome R) has been developed to minimize side effects and increase treatment efficacy [10, 14, 30, 31]; however, the high costs of this drug limited its use. Resistance to AMB is rare, despite 50 years of clinical use as monotherapy, although resistant *C. albicans* strains have been found in different studies [32–35]. The alterations in the composition of the sterols and phospholipids of the membrane, the regulation of oxidative stress, and alterations of the fungal cell are the more frequent resistance mechanisms described for AMB in fungi [10, 12, 14]. In *C. albicans*, resistance to AMB is associated with ergosterol replacement by a precursor molecule or by sterols reduction at the plasma membrane (**Figure 2**) [10, 12, 14].

#### *2.2.1 Alteration in the composition of sterols in the cell membrane (mutations in ERG genes)*

The most common mechanism for acquired resistance to AMB in *C. albicans* is attributed to alterations in the composition of sterols of the fungal cell membrane [10, 12, 14, 36]. Different mutations in *ERG* genes (*ERG11, ERG3, ERG2,* and

*ERG6*) have been associated with this mechanism in *Candida* spp. [14, 37, 38]. Loss of function of the *ERG11* and *ERG3* genes (lanosterol 14α-demethylase and C-5 sterol desaturase, respectively), leads to the exchange of ergosterol for alternative sterols such as lanosterol, eburicol, and 4,14-dimethyl-zymosterol in the membrane of *C. albicans*, [14, 36, 39]. Resistance to AMB in *C. albicans* is also associated with an aminoacidic substitution in *ERG11* and with *ERG5* (sterol desaturase C-22) disfunction, again associated with an alternative membrane sterol composition [14, 39, 40]. In other *Candida* spp., the inactivation of *ERG6* [11, 14, 37] and *ERG2* had a similar effect [11, 14]. Resistance to AMB is rarely found in combination with resistance to other antifungal drugs, although certain mutations that induce resistance to polyenes can lead to cross-resistance to azoles [14, 36, 41].

#### *2.2.2 Response to oxidative stress and alterations in the cell wall*

Fungal resistance mechanisms are also related to oxidative stress regulation, allowing the cell to tolerate exposure to AMB [14, 30]. In *C. albicans*, one of the described mechanisms of stress tolerance to AMB includes the heat shock protein 90 (Hsp90) molecular chaperone, which regulates a large number of proteins involved in several fungal cellular processes [42, 43]. In addition to alterations in the composition of sterols in the plasma membrane and the regulation of oxidative stress, studies in fungi have correlated resistance to AMB with fungal cell wall alterations [14, 44, 45]. In AMB resistant *C. tropicalis* strains, an enlargement of the cell wall has been observed with increased levels of 1,3-β-glucans [14, 44], suggesting an affectation in the penetration of AMB through the cell wall [14, 45].

### **2.3 Echinocandins**

Echinocandins are lipo-peptides that inhibit 1,3-β-d-glucan synthetase, which is responsible for the biosynthesis of 1,3-β-d-glucan, one of the main components of the fungal cell wall, causing osmotic instability and therefore the death of fungal cells (**Figure 1**) [10, 13]. This class of drugs has certain advantages attributable to its effects on the fungal cell wall, including a lower risk of side effects since animal cells do not have this structure [10]. Echinocandins have a limited spectrum, but for *Candida* species, they have broad fungicidal activity. The 1,3-β-d-glucan synthetase target comprises a GTP binding protein Rho, which helps regulate the biosynthetic capacity of glucan synthetase, and a catalytic subunit, FKS, which encodes three related genes, *FKS1*, *FKS2*, and *FKS3*. *FKS1* is essential in *C. albicans* and other *Candida* spp. Whereas *FKS1* and *FKS2* are functionally redundant in *C. glabrata*, *FKS3* is very low expressed compared to other genes [46], not being a significant contributor to biosynthetic capacity in general.

Echinocandins are the first major new class of antifungal drugs on the market in decades. Consequently, it is of vital importance to assess the nature of the resistance mechanism to this class of drugs. Mutations that affect the target site are the most likely resistance mechanism that exists (**Figure 2**), since unlike azoles, echinocandins are poor substrates for drug exit through efflux transporters, ruling out this mechanism of resistance [10, 13]. Specific mutations have already been reported in two highly conserved regions of the Fks1 subunit of glucan synthetase, a membrane protein, which can confer resistance *in vitro* in *Candida* isolates to caspofungin, the first echinocandin approved for the treatment of yeast infections [10, 13, 47, 48]. Other ways in which there may be the acquisition of resistance to echinocandins in *C. albicans* is through different response pathways to cellular stress, as well as some clinical factors such as empirical therapy, prophylaxis, gastrointestinal reservoirs, or intra-abdominal infections.

*2.3.1 Acquired FKS mutations*

Resistance-associated amino acid substitutions occur in two highly conserved hot-spot (HS) regions of the FKS genes. The residues they encompass are Phe641– Pro649 and Arg1361 in *C. albicans* and other *Candida* spp. Substitutions of amino acids Ser645 and Phe641 cause 75% resistance in *C. albicans* [10, 13]. Pharmacodynamic studies conducted in murine models infected with *C. albicans* demonstrated that mutations in the *FKS1* gene confer resistance to echinocandins [48, 49]. Mutations in *FKS1* lead to a decrease in the virulence of *C. albicans* in murine models of IC. Furthermore, high doses of caspofungin are effective against *C. albicans*, including resistant isolates that presented point mutations in *FKS1* [50, 51]. Several studies have reported that mutations in the *FKS1* gene produce changes in the morphology of the cell wall of *C. albicans*, observing a decrease in 1,3-β-d-glucan levels in contrast to the increased amount of chitin in response to echinocandin exposure [51]. Data suggest that increased chitin in the *C. albicans* cell wall could provide a window of opportunity to acquire mutations in *FKS1*, even without exposure to caspofungin [52].

*2.3.2 Adaptive stress responses*

The fungal cell wall is a dynamic structure that changes during growth and development, requires 1,3-β-d-glucan crosslinking, an essential polymer for the survival of the fungal cell. Echinocandins alter the integrity of the cell wall and induce stress in the cell. In response to this, the fungal cell possesses a repertoire of mechanisms to protect the cell against such destabilization. Protection against cell wall weakening is induced through a variety of stress adaptation mechanisms, which involve protein kinase C (PKC), calcineurin, and Hsp90 [10, 13]. Stress signals in the cell wall are transmitted through the Rho GTPase, which mobilizes various effectors. Its activation alters several carbohydrate polymers along with the structure and remodelling of the cell wall. The Hsp90 heat shock protein organizes a cellular stress response circuit that has a major impact on resistance to echinocandins. Also, the genetic or chemical modulation of the Hsp90 protein reduces tolerance to echinocandins [52]. In response to the inhibition of FKS by the action of echinocandins, a greater amount of chitin is produced helping to maintain the integrity of the cell wall as chitin replaces 1,3-β-d-glucan, thus reducing sensitivity to drugs [10, 13, 48].

## 3. New antifungals

The resistance of *C. albicans* and other pathogenic fungi to current antifungal agents has established the need to find new antifungal targets with a novel mechanism of action. Resistant strains are increasing in number for some classes of antifungal agents, particularly for azoles and echinocandins [53]. Consequently, it is necessary to face the challenge of successfully managing fungal infections. To achieve this, one of the main points is the continuation of the development of new antifungal drugs [54]. The main issues faced by the development of new drugs are: 1) they must have a broad spectrum against emerging filamentous yeasts and fungi and 2) they must have a more efficient fungicidal activity to eliminate pathogens quickly and totally [55–59]. Besides, invasive candidiasis occurs in very frail patients who do not tolerate much organ toxicity, since such patients are often taking many other therapeutic agents, so drug–drug interactions must be carefully considered [60].

### 3.1 Discovery and development of new antifungal drugs

This part of the chapter provides an overview of ongoing efforts to develop new classes of antifungal drugs (**Table 1**). Although there are several strategies for the development of these drugs, these include those obtained from new chemical agents, from reusing existing drugs, from peptides with antimicrobial properties, and finally from natural compounds extracted from plants [10, 55, 58].

Several new chemical-antifungals are designed specifically to target either 1,3-β-d-glucan (such as Rezafungin and Ibrexafungerp) or ergosterol (such as the compound VT-1161). These compounds are very specific for fungal infections or they have a longer half-life, offering better efficacy [58, 60–62]. At the same time, several of these antifungal agents have new targets and subsequently, new mechanisms of action. For instance, fosmanogepix, formerly APX001, and aureobasidin A, which act by inhibiting inositol acyltransferase, and inositol phosphorylceramide synthase, respectively [63, 64]. Efungumab (or Mycograb) and geldanamycin-like agents can inhibit the HSP90 chaperone, which has been also shown to confer resistance to antifungals [65, 66]. The AR-12 compound deregulates chaperone's activity by blocking fungal acetyl-CoA synthase [67]. The T-2307 compound is an arylamidine that inhibits the respiratory chain complex and is active against yeast and filamentous fungi [68]. Finally, the VL-2397 compound has a similar structure to the ferrichrome siderophore, and whose mechanism of action or its target is unknown, but it is known to be transported by the Sit1 protein [69]. Some compounds that have been already tested for other types of diseases are now receiving a new focus as antifungals. These include two compounds that enhance the antifungal activity, such as rifampin, which acts on RNA polymerase [70], and verapamil, which acts on a calcium channel [71]. We have also given importance to alternative compounds such as peptides and plant extracts; many molecules are actually studied with promising results, especially against *C. albicans*. Some peptides such as lysozyme, lactoferrin, defensins, Histatin-5, and cathelicidins are known to have antifungal properties. The main mechanism of action is due to the enhancement of substances traffic through the fungal membrane, which favours permeabilization [10, 72–76]. Plant extracts are another prominent source of new antifungals, they can act either alone or synergistically with existing antifungals to improve their function. The compounds extracted from plants are essential oils, terpenes, and flavonoids among many others. They have diverse mechanisms of action, such as alteration of the plasma membrane, binding to ergosterol, induction of apoptosis, inhibition of growth, filamentation, and biofilm formation in *C. albicans* [10, 77–81].

### 3.2 New targets and alternative approaches

Despite the efforts made to discover, repositioning, or create new antifungal drugs, it is imperative to find new targets to help eliminating *Candida* spp. infection. The new antifungal targets include biosynthetic and signal transduction pathways, which are key players for fungal survival processes. The sphingolipids biosynthesis is a biosynthetic pathway considered as a promising target. Sphingolipids are a part of cell membranes, that act as signalling molecules regulating processes such as apoptosis. As fungal sphingolipids are structurally different to mammalian sphingolipids, they are excellent candidates for antifungal design [82]. The heat-shock proteins (Hsps) represent another potential antifungal target as they control several basic physiological activities, and heat-shock protein disruption in *C. albicans* inhibits growth or reverses tolerance to antifungals [83]. A recently studied pathway as a potential target is the ionic homeostasis signalling pathway, which is central to the fungus survival by regulating gene expression,

| Source | Compound | Target | Mechanism of action | Reference |
|---|---|---|---|---|
| **Chemicals** | Rezafungin (CD101) | β-d-glucan | β-d-glucan synthase inhibition | [60] |
| | Ibrexafungerp (SCY-078) | β-d-glucan | β-glucan synthase inhibition | [61] |
| | VT-1161 | Ergosterol | Specific for fungal Cyp51 | [62] |
| | Fosmanogepix (APX001] | Glycosyl phosphatidylinositol | GPI biosynthesis inhibition | [63] |
| | Aureobasidin A | Inositol phosphorylceramide synthase | Sphingolipids biosynthesis inhibition | [64] |
| | Efungumab (or Mycograb) | HSP90 | Antibody binds to fungal HSP90 | [65] |
| | Geldanamycin-like agents | HSP90 | HSP90 inhibition | [66] |
| | AR-12 | Probably blocks fungal acetyl-CoA synthetase 1 | Downregulation of chaperone proteins | [67] |
| | T – 2307 | Mitochondrial membrane potential | Respiratory chain complexes inhibition | [68] |
| | VL-2397 (ASP2397) | Unknown | Unknown, but taken up by Sitl | [69] |
| **Repurposed compounds** | Rifampin | RNA polymerase | Enhance the antifungal activity | [70] |
| | Verapamil | Calcium channel | Enhance the antifungal activity | [71] |
| **Promising Peptides** | Lysozyme | Secreted aspartic protease (SAP) | Reduces SAP activity and secretion | [72] |
| | Lactoferrin (hl.f) | Antimicrobial activity | Production of cationic antimicrobial peptide lactoferricin | [73] |
| | Human b-defensins (HBD) | Cell membrane | Increases membrane permeability | [74] |
| | Histatin-5 | Non-lytic ATP efflux | Inhibition of adhesion | [75] |
| | Cathelicidins | Cell membrane | Increases membrane permeability | [76] |
| | *Scutellaria aicalensis* (flavonoid baicalein) | Unknown | Induces apoptosis in *C. albicans* | [77] |
| | *Cymbopogon nardus* (essential oils) | Unknown | Inhibits hyphal growth in *C. albicans* | [78] |

| Source | Compound | Target | Mechanism of action | Reference |
|---|---|---|---|---|
| **Plant** | *Artemisia judaica* (essential oil) | Germination | Inhibits the formation of germination tube and biofilms in *C. albicans* | [79] |
| **Natural compounds** | Thymol (terpene) | Ergosterol | Binds to ergosterol in the membrane resulting in cell death | [80] |
| | Carvacrol (terpene) | Cell membrane | Alters cellular cytoplasmic membrane and induces apoptosis | [81] |

**Table 1.**
*Antifungal compounds in development against* C. albicans *or* Candida *spp.*

morphological transition, response to stress, and resistance to antifungals [84]. The Ras-cAMP-PKA signal transduction pathway is essential for cellular metabolism and controls morphogenesis, adhesion, and biofilm formation, making the inactivation of this signalling cascade attractive as a target for new antifungals [85].

Finally, an alternative approach to conventional antifungal drugs is the use of nanotechnology, which produces the so-called "nanoantibiotics". These nanoantibiotics are unique due to their improved physicochemical properties, such as reduced toxicity and biocompatibility as well as their size that must be less than 100 nm [86]. The antimicrobial properties of silver have been known for a long time, so silver nanoparticles were tested as antimicrobials and showed potent activity against drug-resistant fungal biofilms [87].

## 4. Conclusions

A better understanding of the resistance mechanisms of azoles, polyenes, and echinocandins, along with the discovery of new cellular and clinical factors promoting resistance, will facilitate the design of more effective strategies to overcome and prevent resistance to antifungal agents. Even though several biomedical research offer a window hoping to reduce the incidence of *C. albicans* and the complications those systemic infections by this fungus entail; the quest for new targets with novel mechanisms of action continues to be the priority.

## Acknowledgements

ALMR thanks the National Council of Science and Technology of Mexico (CONACyT) for the postdoctoral fellowship granted. RIAS and KCP are thankful for the scholarship granted by the National Council of Science and Technology of Mexico (CONACyT).

## Conflict of interest

All authors declare no conflicting interests.

## Author details

Estela Ruiz-Baca*, Rosa Isela Arredondo-Sánchez, Karina Corral-Pérez, Angélica López-Rodríguez, Iván Meneses-Morales, Víctor M. Ayala-García and Ana Lilia Martínez-Rocha*
Faculty of Chemistry Science, Juarez University of Durango State, Durango, México

*Address all correspondence to: eruiz@ujed.mx and analilia.martinez@ujed.mx

## References

[1] Rodriguez DL, Quail MM, Hernday AD, Nobile CJ. Transcriptional circuits regulating developmental processes in *Candida albicans.* Frontiers in Cellular and Infection Microbiology. 2020; **10**:605711.

[2] Maheronnaghsh M, Fatahinia M, Dehghan P, Teimoori A. Identification of *Candida* species and antifungal susceptibility in cancer patients with oral lesions in Ahvaz, Southern West of Iran. *Advanced Biomedical Research*. 2020; **9**:50.

[3] Kan S, Pang Q, Song N, Mei H, Zheng H, Li D, et al. Study on vulvovaginal candidiasis: clinical epidemiology and *in vitro* susceptibility of pathogenic yeasts in China. Social Science Research Network. 2020; **10**:2139.

[4] Forastiero A, Garcia-Gil V, Rivero-Menendez O, Garcia-Rubio R, Monteiro MC, Alastruey A, Jordan R, et al. Rapid development of *Candida krusei* echinocandin resistance during caspofungin therapy. American Society for Microbiology. 2015; **59**:6975-6982.

[5] Pappas PG, Kauffman CA, Andes DR, Clancy CJ, Marr KA, Ostrosky-Zeichner L, et al. Clinical practice guideline for the management of candidiasis: Update by the Infectious Diseases Society of America. Clinical Infectious Diseases. 2016; **62**: e1–e50.

[6] Espinel-Ingroff A, Cantón E, Pemán J. Antifungal Resistance among Less Prevalent *Candida* Non-*albicans* and Other Yeasts versus Established and under Development Agents: A Literature Review. Journal of Fungi. 2021; **7**(1):24.

[7] Pappas PG, Rex JH, Lee J, Hamill RJ, Larsen RA, Powderly W, et al. A prospective observational study of candidemia: epidemiology, therapy, and influences on mortality in hospitalized adult and pediatric patients. Clinical Infectious Diseases. 2003; 37:634-643.

[8] Ford CB, Funt JM, Abbey D, Issi L, Guiducci C, Martinez DA, et al. The evolution of drug resistance in clinical isolates of *Candida albicans.* eLife. 2015; **4**: e00662.

[9] Ruiz CI, Cuenca EM. Antifungals for systemic use. Enfermedades Infecciosas y Microbiología Clínica. 2009; 27(6):353-362.

[10] de Oliveira Santos GC, Vasconcelos CC, Lopes AJO, Cartágenes MSS, Filho AKDB, Nascimento FRF, et al. *Candida* infections and therapeutic strategies: Mechanisms of action for traditional and alternative agents. Frontiers in Microbiology. 2018; **9**(1351):1-25.

[11] Ahmad S, Joseph L, Parker JE, Asadzadeh M, Kelly SL, Meis JF et al. ERG6 and ERG2 are major targets conferring reduced susceptibility to amphotericin B in clinical *Candida glabrata* isolates in Kuwait. Antimicrobial Agents and Chemotherapy. 2019; **63**.

[12] Cuenca EM. Antifungal drug resistance mechanisms in pathogenic fungi: from bench to bedside. Clinical Microbiology and Infection. 2014; 20:54-59.

[13] Houšt' J, Spížek J, Havlíček V. Antifungal Drugs. Metabolites. 2020; **10**(106): 1-16.

[14] Carolus H, Pierson S, Lagrou K, Van Dijck P. Amphotericin B and other polyenes-discovery, clinical use, mode of action and drug resistance. Journal of Fungi (Basel). 2020; **6**(4):321.

[15] Patil A, Majumdar S. Echinocandins in antifungal pharmacotherapy. Journal

of Pharmacy and Pharmacology. 2017; **69**:1635-1660.

[16] Jensen RH, Thyssen KM, Vale L, Sanglard D, Jorgensen R, Fog K, et al. Stepwise emergence of azole, echinocandin and amphotericin B multidrug resistance *in vivo* in *Candida albicans* orchestrated by multiple genetic alterations. Journal of Antimicrobial Chemotherapy. 2015; **70**:2551-2555.

[17] Campoy S, Adrio JL. Antifungals. Biochemical Pharmacology. 2016.

[18] Sanglard D, Coste AT. Activity of isavuconazole and other azoles against *Candida* clinical isolates and yeast model systems with known azole resistance mechanisms. Antimicrob Agents Chemother. 2016; **60**(1):229-238.

[19] Shukla PK, Singh P, Yadav RK, Pandey S, Bhunia SS. Past, present, and future of antifungal drug development. Communicable Diseases of the Developing World. 2016; 125-167.

[20] Cowen LE, Sanglard D, Howard SJ, Rogers PD, Perlin DS. Mechanisms of antifungal drug resistance. Cold Spring Harbor Perspectives in Medicine. 2015; **5**.

[21] Mishra NN, Prasad T, Sharma N, Payasi A, Prasad R, Gupta DK, et al. Pathogenicity and drug resistance in *Candida albicans* and other yeast species a review. Acta Microbiologica et Immunologica Hungarica. 2007; **54**(3):201-235.

[22] Xiang MJ, Liu JY, Ni PH, Wang S, Shi C, Wei B, et al. ERG11 mutations associated with azole resistance in clinical isolates of *Candida albicans*. Federation of European Microbiological Societies. 2013; **13**:386-393.

[23] Vandeputte P, Ferrari S, Coste AT. Antifungal resistance and new strategies to control fungal infections. International Journal of Microbiology. 2012.

[24] Martel CM, Parker JE, Bader O, Weig M, Gross U, Warrilow AG, et al. A clinical isolate of *Candida albicans* with mutations in ERG11 (encoding sterol 14α-demethylase) and ERG5 (encoding C22 desaturase) is cross resistant to azoles and amphotericin B. Antimicrobial Agents and Chemotherapy. 2010; **54**:3578-3583.

[25] Morio F, Pagniez F, Lacroix C, Miegeville M, Pape PL. Amino acid substitutions in the *Candida albicans* sterol D5,6-desaturase (ERG3p) confer azole resistance: characterization of two novel mutants with impaired virulence. Journal of Antimicrobial Chemotherapy. 2012; **67**:2131-2138.

[26] Cannon RD, Lamping E, Holmes AR, Niimi K, Baret PV, Keniya M, et al. Efflux-mediated antifungal drug resistance. Clinical Microbiology Reviews. 2009; **22**(2): 291-321.

[27] Kinsky SC. Polyene antibiotics. In Antibiotics, Springer: 1967; pp. 122-141

[28] Delattin N, Cammue BP, Thevissen K. Reactive oxygen species-inducing antifungal agents and their activity against fungal biofilms. Future Medicinal Chemistry. 2014; **6**(1):77-90.

[29] Darisipudi MN, Allam R, Rupanagudi KV. Polyene macrolide antifungal drugs trigger interleukin-1 b secretion by activating the NLRP3 inflammasome. PLOS one. 2011; **6**(5):1-6.

[30] Mesa-Arango AC, Scorzoni, L, Zaragoza O. It only takes one to do many jobs: Amphotericin B as antifungal and immunomodulatory drug. Frontiers in Microbiology. 2012; 3.

[31] Nett JE, Andes DR. Antifungal agents: spectrum of activity, pharmacology, and clinical indications.

Infectious Disease Clinics of North America. 2016; **30**:51-83.

[32] Rambach G, Oberhauser H, Speth C, Lass CF. Susceptibility of *Candida* species and various moulds to antimycotic drugs: use of epidemiological cutoff values according to EUCAST and CLSI in an 8-year survey. Medical Mycology. 2011; **49**:856-863.

[33] Ostrosky LZ, Rex JH, Pappas PG, Hamill RJ, Larsen RA, Horowitz HW, et al. Antifungal susceptibility survey of 2,000 bloodstream *Candida* isolates in the United States. Antimicrobial Agents and Chemotherapy. 2003; **47**:3149-3154.

[34] Maraki S, Mavromanolaki VE, Stafylaki D, Nioti E, Hamilos G, Kasimati A. Epidemiology and antifungal susceptibility patterns of *Candida* isolates from Greek women with vulvovaginal candidiasis. Mycoses. 2019; **62**:692-697.

[35] Badiee P, Alborzi A. Susceptibility of clinical *Candida* species isolates to antifungal agents by E-test, Southern Iran: A five-year study. Iranian Journal of Microbiology. 2011; **3**:183-188.

[36] Sanglard D, Ischer F, Parkinson T, Falconer D, Bille J. *Candida albicans* mutations in the ergosterol biosynthetic pathway and resistance to several antifungal agents. Antimicrobial Agents and Chemotherapy. 2003; **47**:2404-2412.

[37] Young LY, Hull CM, Heitman J. Disruption of ergosterol biosynthesis confers resistance to amphotericin B in *Candida lusitaniae*. Antimicrobial Agents and Chemotherapy. 2003; **47**:2717-2724.

[38] Silva LN, Oliveira SS, Magalhães LB, Andrade VV, Torres-Santos EC, Carvalho MD, Pereira MD, Branquinha MH, Santos AL. Unmasking the amphotericin B resistance mechanisms in *Candida haemulonii* species complex. ACS Infectious Diseases. 2020; **6**:1273-1282.

[39] Vincent BM, Lancaster AK, Scherz RS, Whitesell L, Lindquist S. Fitness trade-offs restrict the evolution of resistance to amphotericin B. PLOS Biology. 2013; **11**: e1001692.

[40] Martel CM, Parker JE, Bader O, Weig M, Gross U, Warrilow AG, et al. Identification and characterization of four azole-resistant ERG3 mutants of *Candida albicans*. Antimicrobial Agents and Chemotherapy. 2010; **54**(11):4527-4533.

[41] Kelly S, Lamb D, Kelly D, Manning N, Loeffler J, Hebart H, et al. Resistance to fluconazole and cross-resistance to amphotericin B in *Candida albicans* from AIDS patients caused by defective sterol Δ5, 6-desaturation. FEBS letters. 1997; **400**:80-82.

[42] LaFayette SL, Collins C, Zaas AK, Schell WA, Betancourt-Quiroz M, Gunatilaka AL, et al. PKC signaling regulates drug resistance of the fungal pathogen *Candida albicans* via circuitry comprised of Mkc1, calcineurin, and Hsp90. PLOS Pathog. 2010; **6**:e1001069.

[43] Cowen LE, Lindquist S. Hsp90 potentiates the rapid evolution of new traits: drug resistance in diverse fungi. Science. 2005; **309**:2185-2189.

[44] Mesa-Arango AC, Rueda C, Román E, Quintin J, Terrón MC, Luque D, et al. Cell wall changes in amphotericin B-resistant strains from *Candida tropicalis* and relationship with the immune responses elicited by the host. Antimicrobial Agents and Chemotherapy. 2016; **60**:2326-2335.

[45] Seo K, Akiyoshi H, Ohnishi Y. Alteration of cell wall composition leads to amphotericin B resistance in *Aspergillus flavus*. Microbiology and Immunology. 1999; **43**:1017-1025.

[46] Katiyar SK, Alastruey-izquierdo A, Healey KR, Jhonson ME, Perlin DS, Edlind TD. Fks1 and Fks2 are functionally redundant but differentially regulated in *Candida glabrata*: implications for echinocandin resistance. Antimicrobial Agents and Chemotherapy. 2012; **56**(12):6304-6309.

[47] Balashov SV, Park S, Perlin DS. Assessing Resistance to the echinocandin antifungal drug caspofungin in *Candida albicans* by profiling mutations in FKS1. Antimicrobial Agents and Chemotherapy. 2006; **50**(6):2058-2063.

[48] Perlin DS. Mechanisms of echinocandin antifungal drug resistance. Annals of the New York Academy of Science. 2015; **1354** (Pt 1): 1-11.

[49] Garcia-Effron G, Lee S, Park S, Cleary JD, Perlin DS. Effect of *Candida glabrata* FKS1 and FKS2 mutations on echinocandin sensitivity and kinetics of 1,3-beta-D-glucan synthase: implication for the existing susceptibility breakpoint. Antimicrobial Agents and Chemotherapy. 2009; **53** (Pt 9):3690-3699.

[50] Ben-Ami R, García-Effron G, Lewis R, Gamarra S, Leventakos K, Perlin D, et al. Fitness and virulence costs of *Candida albicans* FKS1 hot spot mutations associated with echinocandin resistance. The Journal of Infectious Diseases. 2011; **204**:626-635.

[51] Wiederhold NP, Najvar LK, Bocanegra RA, Kirkpatrick WR, Patterson TF. Caspofungin dose escalation for invasive candidiasis due to resistant *Candida albicans*. Antimicrobial Agents and Chemotherapy. 2011; **55**(7):3254-3260.

[52] Imtiaz T, Lee KK, Munro CA, Maccallum DM, Shankland GS, Johnson EM, et al. Echinocandin resistance due to simultaneous FKS mutation and increased cell wall chitin in a *Candida albicans* bloodstream isolate following brief exposure to caspofungin. Journal of Medical Microbiology. 2012; **61**(Pt 9):1330-1334.

[53] Singh SD, Robbins N, Zaas AK, Schell WA, Perfect JR, Cowen LE. Hsp90 governs echinocandin resistance in the pathogenic yeast *Candida albicans* via calcineurin. PLOS Pathogens. 2009; **5**(7): e1000532.

[54] Gray KC, Palacios DS, Dailey I, Endo MM, Uno BE, Wilcock BC, et al. Amphotericin primarily kills yeast by simply binding ergosterol. Proceedings of the National Academy of Sciences. 2012; **109**(7):2234-2239.

[55] Perfect JR. The antifungal pipeline: a reality check. Nature Reviews Drug Discovery. 2017;**16**(9):603.

[56] Lockhart SR, Etienne KA, Vallabhaneni S, Farooqi J, Chowdhary A, Govender NP et al. Simultaneous emergence of multidrug-resistant *Candida auris* on 3 continents confirmed by whole-genome sequencing and epidemiological analyses. Clinical Infectious Diseases. 2017; **64**(2):134-140.

[57] Parente-Rocha JA, Bailão AM, Amaral AC, Taborda CP, Paccez JD, Borges CL, et al. Antifungal resistance, metabolic routes as drug targets, and new antifungal agents: an overview about endemic dimorphic fungi. Mediators of Inflammation. 2017.

[58] Wall G, Lopez-Ribot JL. Current antimycotics, new prospects, and future approaches to antifungal therapy. Antibiotics. 2020; **9**(8):445.

[59] Pappas PG, Lionakis MS, Arendrup MC, Ostrosky-Zeichner L, Kullberg BJ. Invasive candidiasis. Nature Reviews Disease Primers. 2018; **4**(1):1-20.

[60] Garcia-Effron G. Rezafungin—mechanisms of action, susceptibility and resistance: similarities and differences with the other echinocandins. Journal of Fungi. 2020; **6**(4):262.

[61] Schell WA, Jones AM, Borroto-Esoda K, Alexander BD. Antifungal activity of SCY-078 and standard antifungal agents against 178 clinical isolates of resistant and susceptible *Candida* species. Antimicrobial Agents and Chemotherapy. 2017; **61**(11).

[62] Warrilow AGS, Hull CM, Parker JE, Garvey EP, Hoekstra WJ, Moore,WR, et al. The clinical candidate VT-1161 is a highly potent inhibitor of *Candida albicans* CYP51 but fails to bind the human enzyme. Antimicrobial Agents and Chemotherapy. 2014; **58**(12):7121-7127.

[63] Berkow EL, Lockhart SR. Activity of novel antifungal compound APX001A against a large collection of *Candida auris*. Journal of Antimicrobial Chemotherapy. 2018; **73**(11):3060-3062.

[64] Takesako k, Kuroda H, Inoue T, Haruna F, Yoshikawa Y, Kato I, et al. Biological properties of aureobasidin A, a cyclic depsipeptide antifungal antibiotic. The Journal of Antibiotics. 1993; **46**(9):1414-1420.

[65] Louie A, Stein DS, Zack JZ, Liu W, Conde H, Fregeau C, et al. Dose range evaluation of Mycograb C28Y variant, a human recombinant antibody fragment to heat shock protein 90, in combination with amphotericin B-desoxycholate for treatment of murine systemic candidiasis. Antimicrobial Agents and Chemotherapy. 2011; **55**(7):3295-3304.

[66] Lamoth, F, Alexander BD, Juvvadi PR, Steinbach WJ. Antifungal activity of compounds targeting the Hsp90-calcineurin pathway against various mould species. Journal Antimicrobial Chemotherapy. 2015; **70**:1408-1411.

[67] Koselny K, Green J, Favazzo L, Glazier VE, DiDone L, Ransford S, Krysan DJ. Antitumor/antifungal celecoxib derivative AR-12 is a non-nucleoside inhibitor of the ANL-family adenylating enzyme acetyl CoA synthetase. ACS Infectious Diseases. 2016; **2**(4):268-280.

[68] Yamashita K, Miyazaki T, Fukuda Y, Mitsuyama J, Saijo T, Shimamura S, et al. The novel arylamidine T-2307 selectively disrupts yeast mitochondrial function by inhibiting respiratory chain complexes. Antimicrobial Agents and Chemotherapy. 2019; **63**(8).

[69] Dietl AM, Misslinger M, Aguiar MM, Ivashov V, Teis D, Pfister J, et al. The siderophore transporter Sit1 determines susceptibility to the antifungal VL-2397. Antimicrobial Agents and Chemotherapy. 2019; **63**(10).

[70] Christenson JC, Shalit I, Welch DF, Guruswamy A, Marks MI. Synergistic action of amphotericin B and rifampin against *Rhizopus* species. Antimicrobial Agents and Chemotherapy. 1987; **31**:1775-1778.

[71] Liu S, Yue L, Gu W, Li X, Zhang L, Sun S. Synergistic effect of fluconazole and calcium channel blockers against resistant *Candida albicans*. PLoS One. 2016; **11**(3): e0150859.

[72] Wu T, Samaranayake LP, Leung WK, Sullivan PA. Inhibition of growth and secreted aspartyl proteinase production in *Candida albicans* by lysozyme. Journal Medical Microbiolology 1999; **48**:721-730.

[73] Samaranayake YH, Samaranayake LP, Wu PC, So M. The antifungal effect of lactoferrin and lysozyme on *Candida krusei* and *Candida albicans*. Journal

of Pathology, Microbiology and Immunology.1997;**105**:875-883.

[74] Krishnakumari V, Rangaraj N, Nagaraj R. Antifungal activities of human beta-defensins HBD-1 to HBD-3 and their C-terminal analogs Phd1 to Phd3. Antimicrobial Agents and Chemotherapy. 2009; **53**:256-260.

[75] Edgerton M, Koshlukova SE. Salivary histatin 5 and its similarities to the other antimicrobial proteins in human saliva. Advances in Dental Research. 2000; **14:**16-21.

[76] Tsai PW, Yang CY, Chang HT, Lan CY. (2011). Human antimicrobial peptide LL-37 inhibits adhesion of *Candida albicans* by interacting with yeast cell-wall carbohydrates. PLoS One. 2011; **6**: e17755.

[77] Serpa R, França EJ, Furlaneto-Maia L, Andrade CG, Diniz A, Furlaneto MC. *In vitro* antifungal activity of the flavonoid baicalein against *Candida* species. Journal Medical Microbiolology 2012; **61:**1704-1708.

[78] De Toledo LG, Ramos MADS, Spósito L, Castilho EM, Pavan FR, Lopes ÉDO, et al. Essential oil of *Cymbopogon nardus* (L.) Rendle: a strategy to combat fungal infections caused by *Candida* species. International Journal of Molecular Sciences. 2016; **17**:E1252.

[79] Köse YB, İşcan G, Göger F, Akalın G, Demirci B, Baser KHC. Chemical composition and biological activity of *Centaurea baseri*: new species from Turkey. Chemistry and Biodiversity. 2016; **13**:1369-1379.

[80] de Castro RD, de Souza TMP, Bezerra LM, Ferreira GL, Costa EM, Cavalcanti A L. Antifungal activity and mode of action of thymol and its synergism with nystatin against *Candida* species involved with infections in the oral cavity: an *in vitro* study. 2015; BMC Complementary Medicine and Therapies. **15:**417.

[81] Dalleau S, Cateau E, Bergès T, Berjeaud JM, Imbert C. *In vitro* activity of terpenes against *Candida* biofilms. International Journal Antimicrobial Agents. 2008; **31:**572-576.

[82] Mota FC, Del Poeta M (2020). Fungal sphingolipids: role in the regulation of virulence and potential as targets for future antifungal therapies. Expert Review of Anti-infective Therapy. 2020.

[83] Gong Y, Li T, Yu C, Sun S. *Candida albicans* Heat shock proteins and Hsps-associated signaling pathways as potential antifungal targets. Frontiers in Cellular and Infection Microbiology. 2017; 7:520.

[84] Li Y, Sun L, Lu C, Gong Y, Li M, Sun S. Promising antifungal targets against *Candida albicans* based on ion homeostasis. Frontiers in Cellular and Infection Microbiology. 2018; **8**:286.

[85] Rajasekharan SK, Kamalanathan C, Ravichandran V, Ray AK, Satish AS, Mohanvel SK. Mannich base limits *Candida albicans* virulence by inactivating Ras-cAMP-PKA pathway. Scientific Reports. 2018; **8**(1):1-9.

[86] Beyth N, Houri-Haddad Y, Domb A, Khan W, Hazan R. Alternative antimicrobial approach: nano-antimicrobial materials. Evidence-based Complementary and Alternative Medicine, 2015.

[87] Lara HH, Romero-Urbina DG, Pierce C, Lopez-Ribot JL, Arellano-Jiménez MJ, Jose-Yacaman M. Effect of silver nanoparticles on *Candida albicans* biofilms: an ultrastructural study. Journal of Nanobiotechnology. 2015; **13**(1):1-12.

Chapter 3

# Metabolic Network Modeling for Rational Drug Design against *Candida albicans*

*Rashi Verma, Dibyabhaba Pradhan, Harpreet Singh, Arun Kumar Jain and Luqman Ahmad Khan*

## Abstract

The growing evidences of *Candida albicans* (*C. albicans*) infections are slowly becoming a threat to public health. Moreover, prevalence of antifungal resistant strains of *C. albicans* has emphasized the need for identification of potent targets for rational drug designing. In this aspect, traditional methods for target identification with validation have been found to be expensive and time-consuming. To overcome the concern, genome scale metabolic model construction provides a promising platform that allows novel target identification in combination with subtractive genome analysis. Thus, the chapter details current advancement in model construction, target identification and validation. In brief, it elucidates the overall strategies of *C. albicans* metabolome draft preparation, gap filling, curation of model, simulation followed by model validation, target identification and host pathogen interaction analysis. Finally, several examples of successful metabolic model construction and their utility in rational drug designing also have been discussed.

**Keywords:** Genome Scale Metabolic Model, Target Identification, Drug Designing, Host-Pathogen Interaction, *In-Silico* Gene Knockout

## 1. Introduction

*Candida albicans* (*C. albicans*) is an opportunistic fungal pathogen that lives in equilibrium with normal microbial flora of healthy individual [1]. As commensal, it colonizes on the mucosal surface of oral, respiratory, gastrointestinal and genito-urinary tract. But on transformation into pathogen, it breaches the protective barrier in imunocompromised patients and cause candidiasis [2, 3]. Over 90% patients of cancer and HIV endure with orophryngeal candidiasis whereas vulvovaginal candidiasis distressed 138 million women per year [4, 5]. Candidemia is the most recurrent nosocomial infection acquiring up to 15% infections of blood with mortality rate from 40 to 70% [6]. Consequently, candidiasis has become the most common fungal infections responsible for increased mortality and morbidity worldwide.

For the treatment of candidiasis, limited number of antifungals has been approved for clinical use. These antifungals categorize into four major classes - azoles, polyenes, echinocandins, and pyrimidine analogs [7]. Azole and echinocandins precisely target the enzymes liable for synthesis of cell membrane or cell wall while polyenes directly bind to membrane proteins that maintain the osmolarity of the cell. In addition, Pyrimidines analogs are the sole antifungals that target the pathogen's genome rather than proteome [8, 9]. Consequently, the antifungals disturb the integrity of cell directly or indirectly which ultimately leads to the death of the pathogen (**Figure 1**).

*C. albicans* still represents itself as an emergent pathogen due to the side effects associated with such as RBCs toxicity, nephrotoxicity, hepatotoxicity, arrhythmias, cardiotoxicity and genitointestinal disturbances [10, 11]. Moreover, the drug resistance has also increased the complexity of the disease. The reason behind resistance is the prolonged or discriminated use of antifungals. The resistance mechanism involves the hyperactivity of efflux pumps, mutation in targeted genes and metabolites bypass [12, 13]. Thus, *C. albicans* have different resistance pattern against the diverse antifungals that lift up the difficulty level during the management of

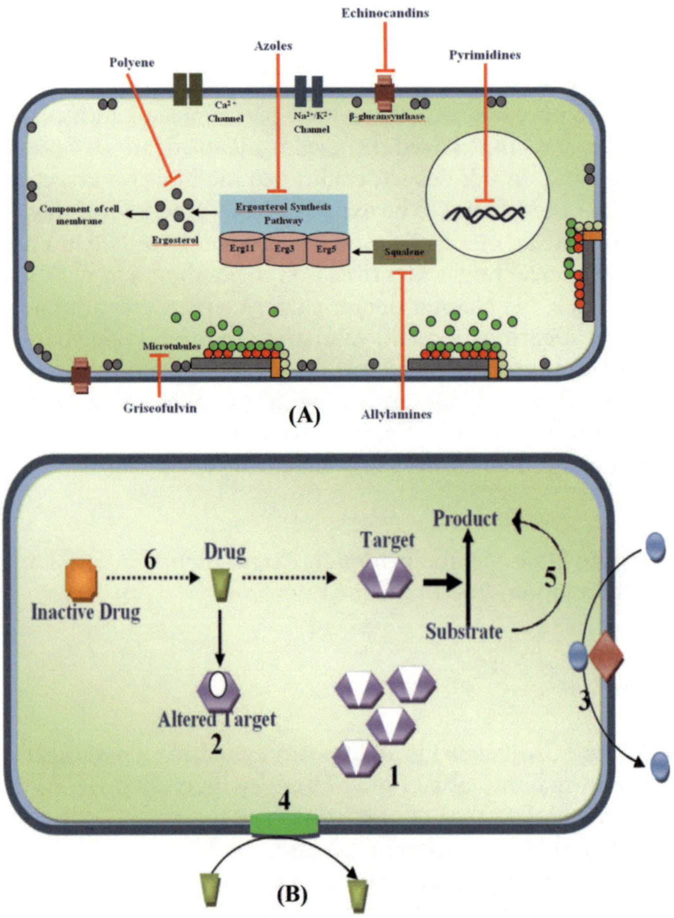


**Figure 1.**

*Antifungal Drug Discovery and Resistance. A) Since 1990s, polyenes, pyrimidine analogs, azoles, echinocandins, and allylamines, morpholines, thiocarbamates has been approved for treatment of* C. albicans *infection. Nystatin and 5-flucytosine binds to membrane ergosterol and thymidylate-synthetase, respectively that leads to the leakage of osmotic constituents. Azoles inhibit the synthesis of ergosterol by interrupting the activity of lanosterol-α-demethylase. Echinocandins halts the participation of (1,3) β-D-glucan synthase in glucan synthesis. Allylamines and thiocarbamates block the oxidation of squalene. Consequently, leads to the death of cell. B) Now, drug resistance has come into picture. Mechanism of resistance involves (1) overexpression of target product, (2) modification of target enzyme, (3) Hyperactivation of multi-drug pump, (4) Production of cell wall Barrier, (5) Adaption to stress response or metabolic bypass, (6) Inactivation of Drug.*

infection. On the other aspect, significant homology of drug targets with human genes/protein, fitness traits and survival strategies such as secretion of hydrolytic enzymes, morphogenetic switch, adhesion to surface and formation of biofilm make this pathogen hard to kill. Thus, the scenario emphasizes the need of the novel drug designing *i.e.* effective against resistant strain of *C. albicans* and easily accessible with less or no side effects [14–18].

Effective drug designing could be possible only after the (i) identification and (ii) validation of potential target. *In silico* and *in vitro* approaches have been attempted for the identification and validation of novel targets of *C. albicans* followed by drug designing. Traditional methods of target identification with experimental validation (growth assay, enzyme inhibition assay, gene knockout, yeast to hybrid system, RNA interference) have been expensive, time-consuming and more focused towards few genes instead of whole genome. To reduce the time and cost, various *in silico* approaches such as subtractive genomics, comparative genomics, machine learning and inverse docking has been performed for novel target identification [19–21]. But a reliable approach is still required for proposed drug target validation.

The present chapter introduces the advancement in reconstruction and analysis of genome scale metabolic model (GSMM) which provides a platform that offers the opportunity to mimic the biological environment of pathogen into a machine to validate the essentiality of the target for the survival of pathogen. "Gene-Protein-Reaction" association in GSMM establishes its importance as a hub for validation of targets while stoichiometry matrix of model helps to depict the linkage information of metabolites to each reaction. If the inhibition of a particular metabolite shows a negative effect on growth of pathogen, it ensures the essentiality of the gene. Additionally, the approach also allows identification of the effect of inhibition on whole metabolome of pathogen. The GSMM can be used independently or in combination with different approaches (high throughput transcriptome profiling and subtractive genomics approach) (**Figure 2**) [19, 20, 22, 23].

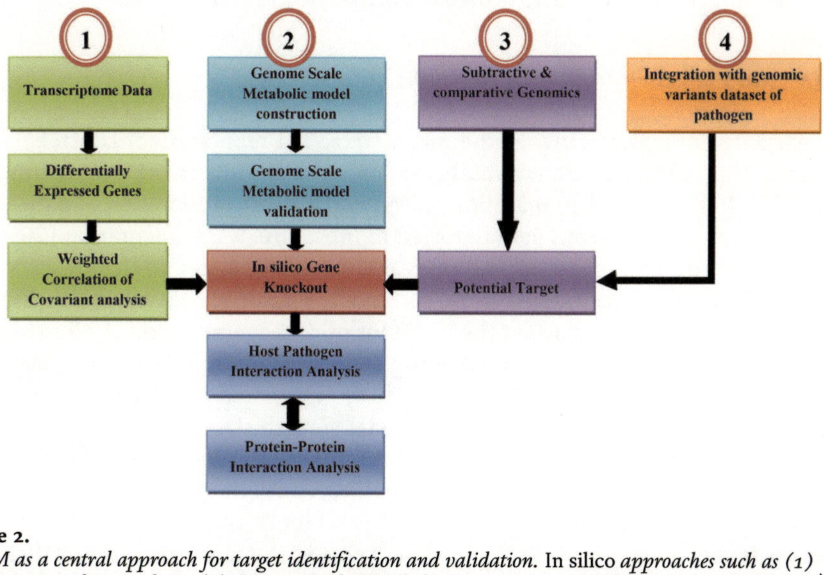


**Figure 2.**
*GSMM as a central approach for target identification and validation.* In silico *approaches such as (1) Transcriptome data analysis, (2) Genome Scale Metabolic Reconstruction (GSMM), (3) Subtractive & Comparative Genomics and (4) Integration of genomics variants dataset of pathogen are widely used for target identification and validation. Among these, GSMM serves as a central approach which can used independently or in combination with these approaches. Gene-protein-reaction association and* in silico *gene knockout feature of GSMM make it reliable and standard approach to validate the putative targets via monitoring the impact of gene deletion on biomass of the system. Further, prioritization of proposed genes can be done by host-pathogen interaction analysis and protein-protein interaction analysis.*

## 2. Genome scale metabolic model reconstruction

A genome scale metabolic model (GSMM) is a computationally designed framework of microorganism that allows an efficient and comprehensive annotation of the metabolic functions of an organism, integrated with large-scale omics datasets and the study of microbe-host interactions [24, 25]. In brief, it describes gene-protein-reaction association of organism that mimics the biological condition in a machine to understand the genetic engineering, protein–protein interaction and evolutionary traits of organism [26]. Consequently, it generates forecast ranging from lethality of pathogen's gene to the dynamics engaged in defense mechanism of host towards infection.

As genome-scale metabolic model reconstruction become the more standard approach, the requirement of *in-silico*, automated tool turn out to be more perceptible to design and analyze these kinds of networks [27, 28]. Furthermore, availability of the whole genome of the pathogen also encourages the construction of *in silico* models. The Recent examples also have shown the potential of these models in the quest for novel drug targets in pathogenic organisms [29–33]. Kim et al., 2009 emerged a model of multi drug resistant *A. baumannii* and find the essential novel targets for therapeutic implications. Abdel-Haleem et al. in 2018, described the reconstruction of genome-scale metabolic models for five life cycle stages of *Plasmodium falciparum*, enabling the identification of potential drug targets that could be used as both, anti-malarial drugs and transmission-blocking agents [34]. Reinksma et al., 2019 developed combine model of *M. tuberculosis* and human to understand the metabolic state of pathogen during infection. Subsequently, Reinksma and team also assessed the effect of increasing dosages of drugs targeting metabolism on the metabolic state of the pathogen and predict resulting metabolic adaptations and flux rerouting through various pathways [35]. Similarly, Nouri et al., 2020 designed a comprehensive model of *Z. mobilis* to find the target for metabolic engineering applications [36]. Thus, design of *C. albicans* would also be the strong platform to understand its metabolic state in distinct adverse conditions that helps to identify and validate the target for novel drug design even against the resistant strains.

## 3. Experimental design for *C. albicans* model

Construction of a model involved 4 major steps: 1) Preparation of Draft; 2) Manual curation; 3) Generation of mathematical model; 4) Network evaluation and analysis [37]. In brief, draft preparation (50%) consist gene annotation of pathogen's genome that further map with data reported in literature. Manual curation (20%) considers the manual refinements and re-evaluation of draft due to the presence of annotations having low confidence score retrieved from organism unspecific biochemical databases that may affect the behavior of pathogenic model. Collection of data for growth condition and biomass composition also is the part of this stage. Generation of mathematical model (10%) is fully automated and includes the conversion of refined draft into mathematical model. Fourth stage comprises the verification, evaluation and validation of model that leads to the identification and fulfillment of network gaps by repeating stages 2 & 3 until the gap fill is accomplished.

The complete protocol of the genome scale metabolic model reconstruction of *C. albicans* is shown in **Figure 3**. The protocol consists of a set of methods that are introduced in sequence but can be combined in a multitude of ways.

### 3.1 Hardware and software

A 64 bit computer of 8 GB RAM with stable internet connection is desired for drafting a model till analysis. MATLAB vR2014b (https://www.mathworks.com/

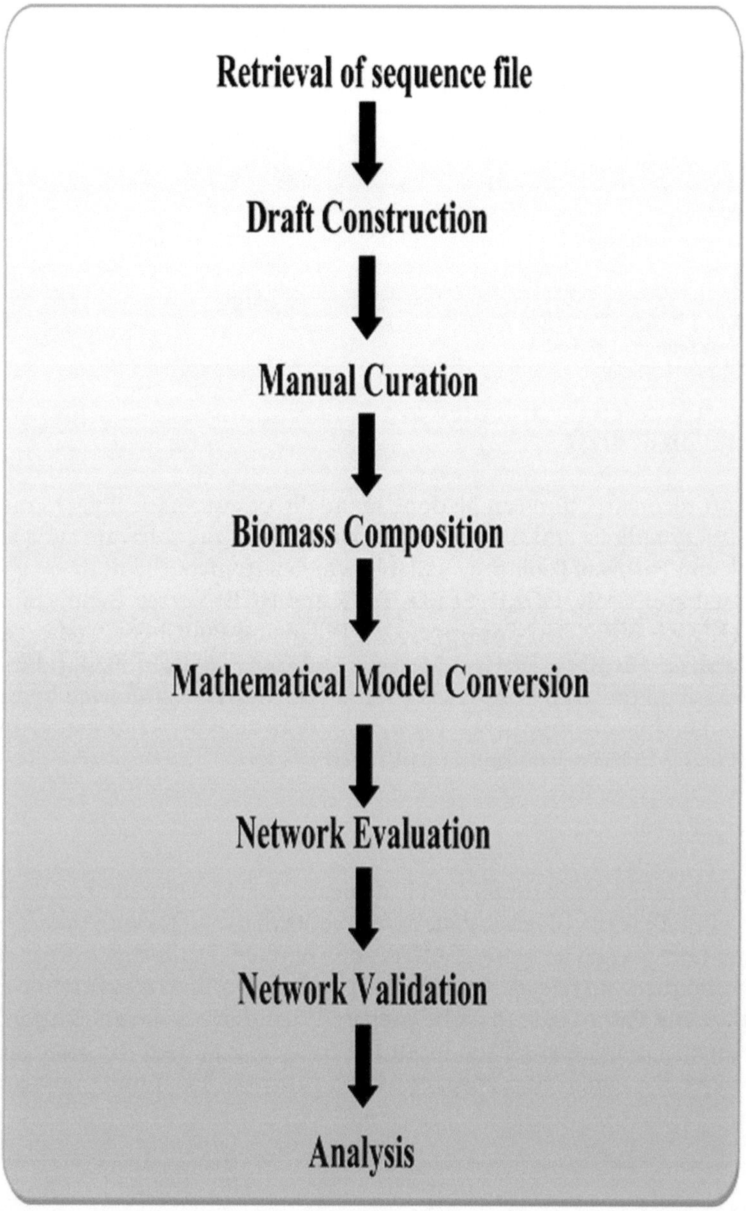


**Figure 3.**
*A flow diagram of Genome Scale Metabolic Model Reconstruction.*

products/matlab.html) or above, COBRA Toolbox v3.0 or above, Pathway Tools v 22.5 are required to accomplished the reconstruction [38–40].

---

**Steps:**

***# Matlab Installation (v2014 or above)***

> > Download → Extract → Click on Setup → Install with or without Internet → Next → Accept license agreement → Next → Provide installation key → Next → Choose Installation Type → Next → Specify installation folder → Next → Provide license file location → Next → Select installation options → Confirm the Installation → Finish

***# CobraTool box Installation***

> > Download i) git ii) curl (v7.0 or above) and iii) CobraTool box (v3.0 or above)

> > First Install git: extract → Click on Setup → choose default settings except adjusting your PATH environment (select use git and optional unix tool from window command prompt) and configuring

---

the line ending conversion (choose checkout as –is, commit Unix-style line ending).
> > Install curl: Select default settings → Just click next → Finish
> > Install CobraTool box: Open git bash → Run command "git clone –depth = 1 https://github.com/opencobra/cobratoolbox.git cobratoolbox" (it will install the setup in C: /user/username/cobratoolbox) → open matlab → click on set path → select the Toolbox folder
***# SBML installation***
> > Download → Extract → Open Matlab → Navigate to SBML toolbox folder → Run script "run (install.m)".
***# Pathway tool Installation***
> > Download → Click on Setup.exe → Select the location of installation (same as cobratool box) → Next → Choose location to store configuration and data file → Next → Verify location of installation → Next → uninstall older version (if present) → Click finish to continue → Ok → Create desktop icon (optional) → Finish

## 3.2 Preparation of draft

The draft reconstruction can be done manually or automatically. On manual mode, it is very tedious and time taking process. Thus, the software such as metaShark and PathwayTools are available which automate the draft by using genome database (CMR, GOLD, SEED, TIGR and NCBI Entrez Gene), biochemical database (KEGG, BRENDA, Transport DB, TCDB and PubChem) and organism-specific database (EcoCyc, BioCyc, Metacyc and Gene Cards) [38, 41]. First, the chapter described the draft construction with PathwayTools followed by manual curation and biomass composition. Further, *in silico* activities and model analysis illustrated using COBRA Toolbox in MATLAB [38–40].

### *3.2.1 Input file format*

PathoLogic plugin of PathwayTool is dedicated for automated draft construction that accepts FASTA file (.fasta), genetic-elements.dat (.dat), GenBank (.gbk) or PathoLogic (.pf) format as input. FASTA and GenBank file formats are easily accessible and can be retrieved from RefSeq and GenBank database while genetic-elements.dat and PathoLogic must be prepared that defines the annotation for each genetic element of organism. Each input file comprises at least the basic attributes such as unique ID, name, start base, end base, function, EC number and gene ontology.

**Steps:**
> > Retrieved the .fasta file and .gbk file of each chromosome of *C. albicans* from RefSeq (https://www.ncbi.nlm.nih.gov/genome/?term=candida+albicans).

### *3.2.2 Creation of new database*

Database creation is the first step of draft model construction that requires the information like unique identifier, database name, taxonomy of organism and database storage type *etc*. Consequently, the provided data is saved into organism.dat and organism-init.dat files that indicate the initialization of new database. Once the database has been initialized, specify the replicons of your organism *i.e.,* the input files of each chromosome that can be .fasta, .dat, .gbk or .pf. Thereafter, specify the reference database of closest organism that will add the missing entities (reactions, enzymes and metabolites) which are absent in databases linked to pathway tool. Trial Parse operation parse the input file(s) to correct the errors present in input file before to automate the building of new database. Finally, the removal of errors allows building the model. This is the focal phase of PathoLogic plugin that perform the parsing again and generate a database for each chromosome, gene, proteins, enzymes, metabolites of organism. Now save the organism database that will take several minutes to complete (**Figure 4, steps: 1–6**).

---

**Steps:**
> > Pathway Tool → PathoLogic (New window popup)
> > Database → Specify Reference PGDB (*S. cerevisiae* or *C. glabrata*)
> > Database → Create New → Organism ID (CanCyc) → NCBI Taxonomy ID (237561)
> > Build → Trial Parse
> > Build → Automated Build.
> > Database → Save KB

---

### *3.2.3 Refinement*

Refining of database includes the inferences and manual operations: **1) Probable Enzymes** involved the additional enzyme-to-reaction assignments; **2) Name Matcher** add the additional name; **3) Rescore Pathways** performs the addition of new pathways and deletion of un-established pathways; **4) Create Protein Complex** permit to stipulate protein complexes that involuntarily link to appropriate reactions; **5) Assign Modified Proteins** allocate the modified substrate within the reaction encoded by gene product within the database; **6) Predict Operons** allow to choose genetic elements on the whole genome; 7) **Transport Inference Parser** finds transport reaction catalyzes proteins to construct their protein complexes and enzymatic reactions; **8) Pathway Hole Filler** seal the gaps using candidate enzymes arise during the construction; **9) Update Cellular Review** draw the cellular outline of database; **10) Consistency Checker** automatically rectify the disturbances of data constraints. Among the refinement, hole filler play the major role which can be done mechanically or manually (**Figure 5A and B**). Now, resave the database and

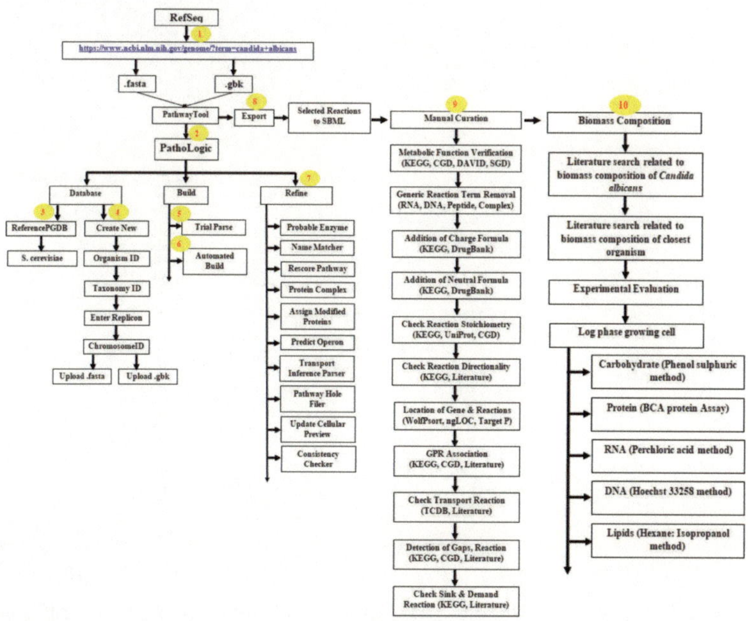


**Figure 4.**
*A detailed Protocol of Genome Scale Metabolic Reconstruction.*

export it for further analysis in .sbml format using the command **File → Export → Selected reactions to SBML file.** For further processing, convert the sbml model into “CanCyc.xls” file for manual curation (**Figure 4, step: 7–8**).

### 3.3 Manual curation

Curation of model is time taking and tedious task that require special consideration during performance. It concentrates on re-evaluation and refinement of model content manually instead of mechanically. The reason behind the manual curation is the absence of proper and complete annotation of gene and their functions. In addition, the available database provides the information which is not organism specific. Consequently, there is a chance of adding those genes or reactions or metabolites which might not be the part of organism’s metabolic network and affect the expected behavior of modeled organism. Thus, it is suggested to curate and assemble the draft model in pathway to pathway manners using KEGG, Gene Ontology, Candida Genome Database, UniProt and DrugBank that ultimately facilitate the detection of gaps of the model [42–45]. Moreover, the stage also includes the metabolic function verification, removal of generic reaction terms (protein, electron acceptor/donor, DNA, RNA *etc.*), addition of charged formula of each metabolite, inspection of reaction stoichiometry as well as directionality, localization of gene with its related reactions, association of gene-protein-reaction, append of transport reaction with literature support. Other than this, inclusion of sink reaction, demand reaction, growth associated and non-growth associated ATP maintenance reaction are also required in a model for *in-silico* growth of the organism (**Figure 4, step: 9**).

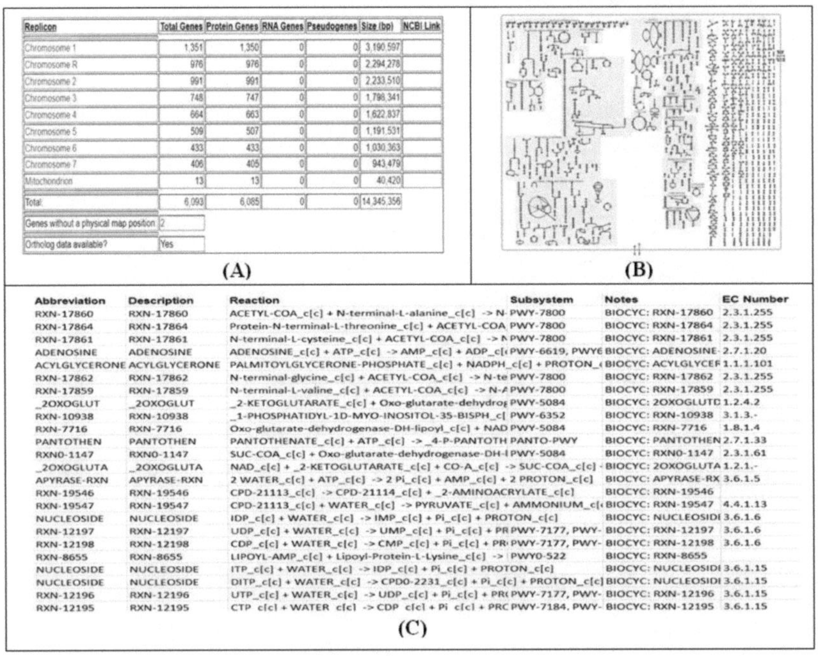

| Replicon | Total Genes | Protein Genes | RNA Genes | Pseudogenes | Size (bp) | NCBI Link |
|---|---|---|---|---|---|---|
| Chromosome 1 | 1,351 | 1,350 | 0 | 0 | 3,190,597 | |
| Chromosome R | 976 | 976 | 0 | 0 | 2,294,278 | |
| Chromosome 2 | 991 | 991 | 0 | 0 | 2,233,510 | |
| Chromosome 3 | 748 | 747 | 0 | 0 | 1,798,341 | |
| Chromosome 4 | 664 | 663 | 0 | 0 | 1,622,837 | |
| Chromosome 5 | 509 | 507 | 0 | 0 | 1,191,531 | |
| Chromosome 6 | 433 | 433 | 0 | 0 | 1,030,363 | |
| Chromosome 7 | 406 | 405 | 0 | 0 | 943,479 | |
| Mitochondrion | 13 | 13 | 0 | 0 | 40,420 | |
| Total: | 6,093 | 6,085 | 0 | 0 | 14,345,356 | |
| Genes without a physical map position | 2 | | | | | |
| Ortholog data available? | Yes | | | | | |

(A)

(B)

| Abbreviation | Description | Reaction | Subsystem | Notes | EC Number |
|---|---|---|---|---|---|
| RXN-17860 | RXN-17860 | ACETYL-COA_c[c] + N-terminal-L-alanine_c[c] -> N- | PWY-7800 | BIOCYC: RXN-17860 | 2.3.1.255 |
| RXN-17864 | RXN-17864 | Protein-N-terminal-L-threonine_c[c] + ACETYL-COA | PWY-7800 | BIOCYC: RXN-17864 | 2.3.1.255 |
| RXN-17861 | RXN-17861 | N-terminal-L-cysteine_c[c] + ACETYL-COA_c[c] -> N | PWY-7800 | BIOCYC: RXN-17861 | 2.3.1.255 |
| ADENOSINE | ADENOSINE | ADENOSINE_c[c] + ATP_c[c] -> AMP_c[c] + ADP_c[ | PWY-6619, PWY6 | BIOCYC: ADENOSINE- | 2.7.1.20 |
| ACYLGLYCERONE | ACYLGLYCERONE | PALMITOYLGLYCERONE-PHOSPHATE_c[c] + NADPH_c[c] + PROTON_( | | BIOCYC: ACYLGLYCEF | 1.1.1.101 |
| RXN-17862 | RXN-17862 | N-terminal-glycine_c[c] + ACETYL-COA_c[c] -> N-te | PWY-7800 | BIOCYC: RXN-17862 | 2.3.1.255 |
| RXN-17859 | RXN-17859 | N-terminal-L-valine_c[c] + ACETYL-COA_c[c] -> N-A | PWY-7800 | BIOCYC: RXN-17859 | 2.3.1.255 |
| _2OXOGLUT | _2OXOGLUT | _2-KETOGLUTARATE_c[c] + Oxo-glutarate-dehydrog | PWY-5084 | BIOCYC: 2OXOGLUTD | 1.2.4.2 |
| RXN-10938 | RXN-10938 | _1-PHOSPHATIDYL-1D-MYO-INOSITOL-35-BISPH_c[ | PWY-6352 | BIOCYC: RXN-10938 | 3.1.3.- |
| RXN-7716 | RXN-7716 | Oxo-glutarate-dehydrogenase-DH-lipoyl_c[c] + NAD | PWY-5084 | BIOCYC: RXN-7716 | 1.8.1.4 |
| PANTOTHEN | PANTOTHEN | PANTOTHENATE_c[c] + ATP_c[c] -> _4-P-PANTOTH | PANTO-PWY | BIOCYC: PANTOTHEN | 2.7.1.33 |
| RXN0-1147 | RXN0-1147 | SUC-COA_c[c] + Oxo-glutarate-dehydrogenase-DH-l | PWY-5084 | BIOCYC: RXN0-1147 | 2.3.1.61 |
| _2OXOGLUTA | _2OXOGLUTA | NAD_c[c] + _2-KETOGLUTARATE_c[c] + CO-A_c[c] -> SUC-COA_c[c] - | | BIOCYC: 2OXOGLUTA | 1.2.1.- |
| APYRASE-RXN | APYRASE-RXN | 2 WATER_c[c] + ATP_c[c] -> 2 Pi_c[c] + AMP_c[c] + 2 PROTON_c[c] | | BIOCYC: APYRASE-RX | 3.6.1.5 |
| RXN-19546 | RXN-19546 | CPD-21113_c[c] -> CPD-21114_c[c] + _2-AMINOACRYLATE_c[c] | | BIOCYC: RXN-19546 | |
| RXN-19547 | RXN-19547 | CPD-21113_c[c] + WATER_c[c] -> PYRUVATE_c[c] + AMMONIUM_c[ | | BIOCYC: RXN-19547 | 4.4.1.13 |
| NUCLEOSIDE | NUCLEOSIDE | IDP_c[c] + WATER_c[c] -> IMP_c[c] + Pi_c[c] + PROTON_c[c] | | BIOCYC: NUCLEOSIDI | 3.6.1.6 |
| RXN-12197 | RXN-12197 | UDP_c[c] + WATER_c[c] -> UMP_c[c] + Pi_c[c] + PR | PWY-7177, PWY- | BIOCYC: RXN-12197 | 3.6.1.6 |
| RXN-12198 | RXN-12198 | CDP_c[c] + WATER_c[c] -> CMP_c[c] + Pi_c[c] + PR | PWY-7177, PWY- | BIOCYC: RXN-12198 | 3.6.1.6 |
| RXN-8655 | RXN-8655 | LIPOYL-AMP_c[c] + Lipoyl-Protein-L-Lysine_c[c] -> | PWY0-522 | BIOCYC: RXN-8655 | |
| NUCLEOSIDE | NUCLEOSIDE | ITP_c[c] + WATER_c[c] -> IDP_c[c] + Pi_c[c] + PROTON_c[c] | | BIOCYC: NUCLEOSIDI | 3.6.1.15 |
| NUCLEOSIDE | NUCLEOSIDE | DITP_c[c] + WATER_c[c] -> CPD0-2231_c[c] + Pi_c[c] + PROTON_c[c] | | BIOCYC: NUCLEOSIDI | 3.6.1.15 |
| RXN-12196 | RXN-12196 | UTP_c[c] + WATER_c[c] -> UDP_c[c] + Pi_c[c] + PR( | PWY-7177, PWY- | BIOCYC: RXN-12196 | 3.6.1.15 |
| RXN-12195 | RXN-12195 | CTP_c[c] + WATER_c[c] -> CDP_c[c] + Pi_c[c] + PRC | PWY-7184, PWY- | BIOCYC: RXN-12195 | 3.6.1.15 |

(C)

**Figure 5.**
*Genome-Scale Metabolic Model of* C. albicans. *(A) Construction of model using Pathway Tools generated the detailed chromosome-wise description and (B) cellular overview of the model that defines the linkage among the metabolic pathways present in model. (C) For manual curaion, sbml format of model is converted to mathematical model which further subjected for evaluation and validation.*

### 3.4 Biomass composition

Biomass reaction is the engine of metabolic model as it shows the obvious effects on model validation and strain improvement. The biomass consist of cellular components (proteins, RNA, DNA, Lipids, Lipopolysaccharides, Peptidoglycan, Glycogen, Polyamines, *etc.*) with its fractional constituents [46]. Estimation of total biomass composition may not be feasible; still the determination of comparative fraction of all precursors can be possible experimentally for log phase growing cells. Among the biomass precursors, lipid extraction is quite tough due the presence of different fatty acid with diverse chain length, saturation and un-saturation. After the quantification of biomass components, data is subjected to normalization that computes the equation of biomass (**Figure 4, step: 10**).

### 3.5 Curated model conversion to mathematical model

In this stage, model is subjected to convert the curated draft into a condition-specific mathematical model *i.e.,* fully automated. MATLAB and COBRA Toolbox are widely used software for model conversion, evaluation and analysis. To convert the model initialize the COBRA toolbox using the command **"initCobraToolbox"** first, then set of optimization solvers such as Gurobi and LP. The optimization solvers provide commanding algorithms that improvise the programming of mathematical models, constraint models and constraint based scheduling models. The solver Gurobi is a default solver for LP, MILP and QP problems while GLPK is selected for LP and MILP problems. Read the model with **"xls2model"** command to verify and set the objective as well as simulation constraints to the model. Save it to "CanCyc.xml" format.

Script to load and save the model in mathematical format is provided in **Supplementary Data 1.**

### 3.6 Model evaluation and validation

The metabolic model designed in third stage may have some common errors: 1) wrong reaction constraints; 2) cofactor cannot be produced or consumed; 3) shuffling of compounds across compartments; 4) missing transport and exchange reactions. To rectify these issues network verification, evaluation and validation is needed. Verification and evaluation usually leads to the addition of transport reaction, exchange reactions and metabolic function that can be done by the repetitive process of stage 2 and 3. Thus, it is also known as iterative process that evokes the debugging to cure errors arising computationally. The major concern is to make a decision when to end this process which is based on the rationale of reconstruction.

The process starts with the test of unbalanced reaction that provides the list of unbalanced reaction in model to balance it manually. Next is to identify the dead end metabolites that are only consumed or produced and indicates about gaps present in the model. Removal of dead metabolites promote gap filling which is a manual process that can be done by using the published literature, genome pathway annotation tools (KEGG) and organism specific databases. During gap fill, all added reactions and metabolites must be connecting to each other. This step also includes the addition of exchange reactions and transport reaction as well. Thereafter, the upper and lower constraints to desire medium or environmental condition required for the growth of organism. Constraints must be varying according to the objective of study.

To practice the exercise of evaluation and validation, user can also download the published GSMM model of other organisms from BioModel Database (http://www.ebi.ac.uk/biomodels/).

---

**Steps:**

```
> > initCobraToolbox;
> > solverok = changeCobraSolver('glpk','LP');
> > model = xls2model(CanCyc.xlsx');
```

**# Check biomass production**

```
> > FBAsolution = optimizeCbModel(model,'max')
```

**# Test unbalances mass and charge of C, N, P and S atoms and balances them.**

```
> > [dE E] = checkBalance(model,'C');
> > ind = find(dE);
> > ImbalReacs_C = model.rxns(ind);
> > [dE E] = checkBalance(model,'N');
> > ind = find(dE);
> > ImbalReacs_N = model.rxns(ind);
> > [dE E] = checkBalance(model,'P');
> > ind = find(dE);
> > ImbalReacs_P = model.rxns(ind);
> > [dE E] = checkBalance(model,'S');
> > ind = find(dE);
> > ImbalReacs_S = model.rxns(ind);
```

**# Identification of metabolic dead ends and document it in excel.**

```
> > model = changeObjective(model,'BiomassReac_1');
> > [missingMets,presentMets] = AnalyzeGaps('model');
```

**# Filling the gap of the model.**

- Search the published literature first on metabolome of *C. albicans*
- Give the second priority to genome annotation tools (KEGG - https://www.kegg.jp/kegg-bin/show_organism?menu_type=pathway_maps&org=cal) and organism specific database (Candida genome database - http://www.candidagenome.org/).

- Last precedence gives to closest organism database (Saccharomyces - https://www.yeastgenome.org/)
- Document all the gap fill reactions with references.
- Add gap fill reaction by repeating the stage 2 till the biomass production occurs. For addition or deletion use the command given below-

```
> > model_add = addReaction(model, 'ReactionName', 'Reaction');
> > model_del = removeRxns(model, 'ReactionName', 'Reaction');
```

- After filling the gaps, check the production of biomass of the cell as commands given below:

```
> > initCobraToolbox;
> > solverok = changeCobraSolver('glpk','LP');
> > model = xls2model(CanCyc.xlsx');
> > model = changeObjective(model,'BiomassReac_1');
> > FBAsolution = optimizeCbModel(model,'max')
```

**# Set the reaction constraints of modeled organism on minimal media**

- For example-check the growth of *C. albicans* at Glucose and oxygen

```
> > model = changeRxnBounds(model, 'GLC_tx_c', −10, 'l');
> > model = changeRxnBounds(model, Oxygen_Molecule_tx_c', −10, 'l');
```

**# Change the objective function according the study and simulate the model**

```
> > model = changeObjective(model,'BiomassReac_2');
> > FBAsolution = optimizeCbModel(model,'max')
```

- Change the reaction and reaction bounds according to the aim of the study and validate the model. 'u' = upper bound, 'l' = lower bound and 'b' = both bound

**# Validation of the model**

- On different carbon sources (Glucose for example).

```
> > GrowthRate = 0.10;
> > CarbonIntake = −10.0;
> > model = changeRxnBounds(model,'BiomassReac_c_test_2',GrwothRate,'b');
> > model = changeRxnBounds(model,'GLC_tx_c',CarbonIntkRate,'b');
> > fba = MinOptmz
```

---

- Similarly perform the validation of model on nitrogen and sulfur sources i.e., reported in the literature. Draw the flux map to check the flow of carbon throughout the model.

Once the model has been developed and validated, write the model in either . mat, .sbml or .xlsx format for further analysis (**Figure 5C**).

### 3.7 Target identification through essentiality analysis and gene knockout

Developed model can be used for novel drug target identification by prioritizing essential genes through *in silico* gene knockout of single or double genes. Reaction deletion command lists out the reactions that help the pathogen model to cultivate too fast. On the other hand, gene knockout inhibit the particular gene of metabolome which act as chokepoint in pathogen. Consequently, it provokes the survival of pathogen that establishes the necessity of that gene as drug target for therapeutic use.

```
# Reaction essentiality analysis
> > [EssentialReactions,NonEssentialReactions,Reactions] = ReacEssentiality(model);
# Gene essentiality analysis
• Single gene deletion
> > [EssentialGenes,NonEssentialGenes,Genes] = SingleGeneEssentiality(model);
• Double gene deletion
> > [EssentialGenes,NonEssentialGenes,Genes] = DoubleGeneEssentiality(model);
# Gene Knockout
• Single gene knockout
> > geneList = "GeneName1";
> > [grRatioDble, grRateKO, grRateWT] = singleGeneDeleion(model, MOMA, geneList);
• Double gene knockout
> > geneList1 = "GeneName1";
> > geneList2 = "GeneName2";
> > [grRatioDble, grRateKO, grRateWT] = doubleGeneDeleion (model, MOMA, geneList1,
geneList2);
```

### 3.8 Host-pathogen interaction (HPI) analysis

HPI analysis comprised of five central stages: 1) reconstruction of high quality host and pathogen model, 2) check the common reaction and metabolites of both the model, 3) integration of the model, 4) integration testing and 5) simulation [47].

HPI model development and analysis is presented in the **Figure 6.** As the protocol for high quality model development has been described above, this section will exclude the first stage i.e., reconstruction of model. Next is pre-integration check to remove the violation of mass conservation. Thus, it checks the mass balance and stoichiometric consistency on prior basis followed by the identification of overlap features of both the models [48]. These regions are expected to be the region of host-pathogen interaction. Thereafter, assign the unique metabolite and reaction identifier to common metabolites and reaction respectively that promotes the integration of both the model successfully.

Once the model has been merged, detailed integration testing required to be carried out that ensure the linkage among the merged models. Integrity testing is divided into functionality test suits and independence suit. Functionality suits include the check mass balance, flux variability analysis and literature based boundaries verification while independence suit needs to find objective function

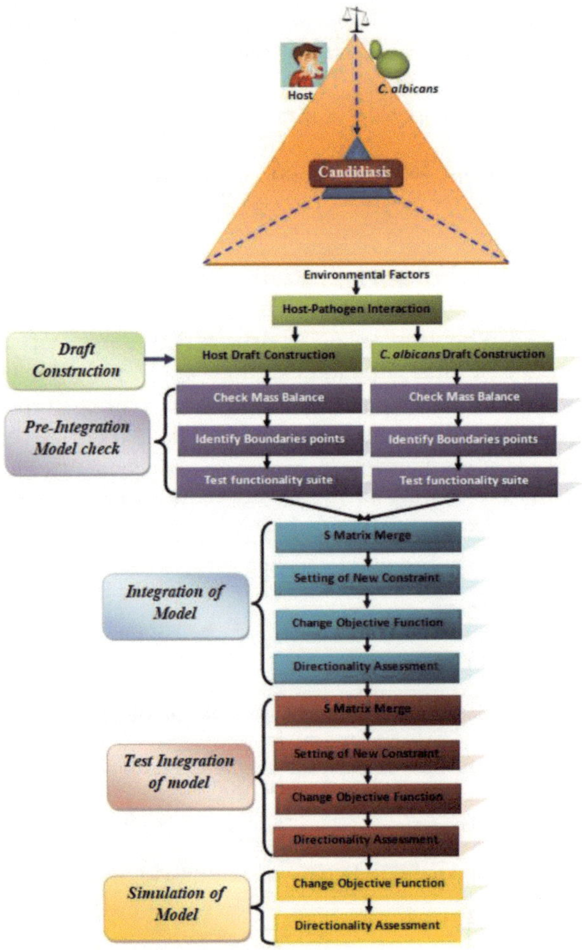


**Figure 6.**
*Host-Pathogen Interaction (HPI) Analysis using GSMM. Host-pathogen interaction analysis using GSMM includes draft construction, pre-integration model check, integration of host-pathogen models, test of integration and simulation. Consequently, the build mathematical model can be used to evaluate the interactions of individual components and highlight potential targets for drug development.*

that assumed to be influence by the host pathogen coupling. Further, the evaluation of function is performed to proceed towards simulation of the integrated model. Through simulation, one can envisage the characteristics of disease or validate the experimental data in infection circumstances using HPI model. In addition, Gene knockout of the HPI model can potentially predict virulent genes of the pathogen with better accuracy than the individual model. The identification of lethal gene and knockout can be performed as similar as mentioned in the section of 3.7.

## 4. Advantages of GSMM of *C. albicans*

Genome Scale metabolic models of several pathogens have been designed and available at Biomodel database. The potential of these models in studying whole metabolite organization in living cell/organism widen significant attention in system medicine. Nevertheless, GSMM of *C. albicans* would provide a platform for target identification and validation. Till date, single metabolic model (iRV781) of *C. albicans* has been developed [49]. The published model, designed on the GUI platform of merlin, found to be non-compatible with widely used system biology platforms such as Matlab and CobraToolbox. On comparison, model of present study explained the complete set of gene-protein-reaction associations based on genome annotation data and experimentally obtained information. Consequently, it allows the production of flux value for entire set of reactions. The model also provides the opportunity to integrate the omics and kinetic data that contributed to better understanding of metabolism of pathogen. Such progression in model development of *C. albicans* permits the context–specific simulation. Additionally, the model would be beneficial in prediction of enzyme functions, pan-reactome analysis, modeling interaction among multiple cells or organism and understanding the colonization of pathogen and disease progression in human. In future prospect, proposed model could be used as a reference template to design the model for resistant strain of *C. albicans.*

## 5. Conclusion

Despite the presence of distinct antifungals, current situation demands the discovery of novel antifungal(s) against resistant strain of *C. albicans*. A successful drug design method is reliable on when a potent target is present. Thus, a potential strategy, approach, pipeline and tools are required to identify the druggable target. System biology and genome scale metabolic reconstruction of infectious pathogen offer a novel and effective approach that positively impel the research towards the identification of drug target that could help to design a novel antifungal against all kind of pathogenic strains of *C. albicans*.

## Acknowledgements

Authors highly acknowledge Indian Council of Medical Research for supporting facilities to carry out the research work in National Institute of Pathology, New Delhi.

## Funding

Authors highly acknowledge the financial assistance provided by Indian Council of Medical Research, India to Ms. Rashi Verma [(ISRM/11(30)/2019) dated 02/09/2019].

## Declaration of interest

No issues of conflicting interest have been declared or identified.

**Author details**

Rashi Verma[1,2], Dibyabhaba Pradhan[3], Harpreet Singh[3], Arun Kumar Jain[2*] and Luqman Ahmad Khan[1*]

1 Department of Biosciences, Jamia Millia Islamia, New Delhi, India

2 Biomedical Informatics Centre, ICMR-National Institute of Pathology, New Delhi, India

3 Computational Genomics Centre, Indian Council of Medical Research, Campus - All India Institute of Medical Sciences, New Delhi, India

*Address all correspondence to: drakjain@gmail.com; lkhan@jmi.ac.in

## References

[1] J. M. Achkar and B. C. Fries, "Candida Infections of the Genitourinary Tract," *Clin. Microbiol. Rev.*, vol. 23, no. 2, pp. 253–273, Apr. 2010, doi: 10.1128/CMR.00076-09.

[2] B. A. Neville, C. d'Enfert, and M.-E. Bougnoux, "*Candida albicans* commensalism in the gastrointestinal tract," *FEMS Yeast Res.*, vol. 15, no. 7, Nov. 2015, doi: 10.1093/femsyr/fov081.

[3] C. J. Nobile and A. D. Johnson, "*Candida albicans* Biofilms and Human Disease," *Annu. Rev. Microbiol.*, vol. 69, pp. 71–92, 2015, doi: 10.1146/annurev-micro-091014-104330.

[4] A. L. Jayachandran *et al.*, "Oral Candidiasis among Cancer Patients Attending a Tertiary Care Hospital in Chennai, South India: An Evaluation of Clinicomycological Association and Antifungal Susceptibility Pattern," *Canadian Journal of Infectious Diseases and Medical Microbiology*, Jun. 14, 2016. https://www.hindawi.com/journals/cjidmm/2016/8758461/ (accessed Dec. 28, 2020).

[5] D. W. Denning, M. Kneale, J. D. Sobel, and R. Rautemaa-Richardson, "Global burden of recurrent vulvovaginal candidiasis: a systematic review," *Lancet Infect. Dis.*, vol. 0, no. 0, Aug. 2018, doi: 10.1016/S1473-3099(18)30103-8.

[6] F. Lamoth, S. R. Lockhart, E. L. Berkow, and T. Calandra, "Changes in the epidemiological landscape of invasive candidiasis," *J. Antimicrob. Chemother.*, vol. 73, no. suppl_1, pp. i4–i13, Jan. 2018, doi: 10.1093/jac/dkx444.

[7] V. Moudgal and J. Sobel, "Antifungals to treat *Candida albicans*," *Expert Opin. Pharmacother.*, vol. 11, no. 12, pp. 2037–2048, Aug. 2010, doi: 10.1517/14656566.2010.493875.

[8] M. A. Ghannoum and L. B. Rice, "Antifungal Agents: Mode of Action, Mechanisms of Resistance, and Correlation of These Mechanisms with Bacterial Resistance," *Clin. Microbiol. Rev.*, vol. 12, no. 4, pp. 501–517, Oct. 1999, doi: 10.1128/CMR.12.4.501.

[9] R. R. Prasad, V. Shree, S. Sagar, S. Kumar, and P. Kumar, "Prevalence and Antifungal Susceptibility of *Candida albicans* in Patna, India," 2016, doi: 10.20546/IJCMAS.2016.504.108.

[10] F. P. Tverdek, D. Kofteridis, and D. P. Kontoyiannis, "Antifungal agents and liver toxicity: a complex interaction," *Expert Rev. Anti Infect. Ther.*, vol. 14, no. 8, pp. 765–776, Aug. 2016, doi: 10.1080/14787210.2016.1199272.

[11] R. E. Lewis, "Current Concepts in Antifungal Pharmacology," *Mayo Clin. Proc.*, vol. 86, no. 8, pp. 805–817, Aug. 2011, doi: 10.4065/mcp.2011.0247.

[12] Z. A. Kanafani and J. R. Perfect, "Antimicrobial resistance: resistance to antifungal agents: mechanisms and clinical impact," *Clin. Infect. Dis. Off. Publ. Infect. Dis. Soc. Am.*, vol. 46, no. 1, pp. 120–128, Jan. 2008, doi: 10.1086/524071.

[13] C. B. Ford *et al.*, "The evolution of drug resistance in clinical isolates of *Candida albicans*," *eLife*, vol. 4, doi: 10.7554/eLife.00662.

[14] J. P. Guirao-Abad, V. Pujante, R. Sánchez-Fresneda, G. Yagüe, and J.-C. Argüelles, "Sensitivity of the *Candida albicans* trehalose-deficient mutants tps1Δ and tps2Δ to amphotericin B and micafungin," *J. Med. Microbiol.*, vol. 68, no. 10, pp. 1479–1488, Oct. 2019, doi: 10.1099/jmm.0.001053.

[15] H. S. Rane *et al.*, "*Candida albicans* Pma1p Contributes to Growth, pH Homeostasis, and Hyphal Formation,"

*Front. Microbiol.*, vol. 10, p. 1012, 2019, doi: 10.3389/fmicb.2019.01012.

[16] S. Zhao *et al.*, "Design, synthesis and evaluation of biphenyl imidazole analogues as potent antifungal agents," *Bioorg. Med. Chem. Lett.*, vol. 29, no. 17, pp. 2448–2451, Sep. 2019, doi: 10.1016/j.bmcl.2019.07.037.

[17] A. T. Sangamwar, U. D. Deshpande, and S. S. Pekamwar, "Antifungals: Need to Search for a New Molecular Target," *Indian J. Pharm. Sci.*, vol. 70, no. 4, pp. 423–430, 2008, doi: 10.4103/0250-474X.44588.

[18] D. J. Krysan, "The unmet clinical need of novel antifungal drugs," *Virulence*, vol. 8, no. 2, pp. 135–137, Jan. 2017, doi: 10.1080/21505594.2016.1276692.

[19] R. Verma, D. Pradhan, M. Maseet, H. Singh, A. K. Jain, and L. A. Khan, "Genome-wide screening and in silico gene knockout to predict potential candidates for drug designing against *Candida albicans*," *Infect. Genet. Evol. J. Mol. Epidemiol. Evol. Genet. Infect. Dis.*, vol. 80, p. 104196, Jun. 2020, doi: 10.1016/j.meegid.2020.104196.

[20] H. Tripathi, S. Luqman, A. Meena, and F. Khan, "Genomic identification of potential targets unique to *Candida albicans* for the discovery of antifungal agents," *Curr. Drug Targets*, vol. 15, no. 1, pp. 136–149, Jan. 2014.

[21] A. Ahmad and A. U. Khan, "Prevalence of Candida species and potential risk factors for vulvovaginal candidiasis in Aligarh, India," *Eur. J. Obstet. Gynecol. Reprod. Biol.*, vol. 144, no. 1, pp. 68–71, May 2009, doi: 10.1016/j.ejogrb.2008.12.020.

[22] S. Choudhary *et al.*, "Transcriptomic landscaping of core genes and pathways of mild and severe psoriasis vulgaris," *Int. J. Mol. Med.*, vol. 47, no. 1, pp. 219–231, Jan. 2021, doi: 10.3892/ijmm.2020.4771.

[23] S. Choudhary, D. Pradhan, N. S. Khan, H. Singh, G. Thomas, and A. K. Jain, "Decoding Psoriasis: Integrated Bioinformatics Approach to Understand Hub Genes and Involved Pathways," *Curr. Pharm. Des.*, vol. 26, no. 29, pp. 3619–3630, Aug. 2020, doi: 10.2174/1381612826666200311130133.

[24] A. Mardinoglu, R. Agren, C. Kampf, A. Asplund, M. Uhlen, and J. Nielsen, "Genome-scale metabolic modelling of hepatocytes reveals serine deficiency in patients with non-alcoholic fatty liver disease," *Nat. Commun.*, vol. 5, p. 3083, 2014, doi: 10.1038/ncomms4083.

[25] A. Heinken and I. Thiele, "Systems biology of host–microbe metabolomics," *WIREs Syst. Biol. Med.*, vol. 7, no. 4, pp. 195–219, 2015, doi: https://doi.org/10.1002/wsbm.1301.

[26] M. A. Oberhardt, B. Ø. Palsson, and J. A. Papin, "Applications of genome-scale metabolic reconstructions," *Mol. Syst. Biol.*, vol. 5, p. 320, 2009, doi: 10.1038/msb.2009.77.

[27] S. N. Mendoza, B. G. Olivier, D. Molenaar, and B. Teusink, "A systematic assessment of current genome-scale metabolic reconstruction tools," *Genome Biol.*, vol. 20, no. 1, p. 158, Aug. 2019, doi: 10.1186/s13059-019-1769-1.

[28] B. K. Chung *et al.*, "Genome-scale metabolic reconstruction and in silico analysis of methylotrophic yeast Pichia pastoris for strain improvement," *Microb. Cell Factories*, vol. 9, p. 50, Jul. 2010, doi: 10.1186/1475-2859-9-50.

[29] A. K. Chavali, K. M. D'Auria, E. L. Hewlett, R. D. Pearson, and J. A. Papin, "A metabolic network approach for the identification and prioritization of antimicrobial drug targets," *Trends Microbiol.*, vol. 20, no. 3, pp. 113–123, Mar. 2012, doi: 10.1016/j.tim.2011.12.004.

[30] T. Ulas, S. A. Riemer, M. Zaparty, B. Siebers, and D. Schomburg, "Genome-Scale Reconstruction and Analysis of the Metabolic Network in the Hyperthermophilic Archaeon Sulfolobus Solfataricus," *PLOS ONE*, vol. 7, no. 8, p. e43401, Aug. 2012, doi: 10.1371/journal.pone.0043401.

[31] I. Larsson, M. Uhlén, C. Zhang, and A. Mardinoglu, "Genome-Scale Metabolic Modeling of Glioblastoma Reveals Promising Targets for Drug Development," *Front. Genet.*, vol. 11, 2020, doi: 10.3389/fgene.2020.00381.

[32] B. S. Mienda, R. Salihu, A. Adamu, and S. Idris, "Genome-scale metabolic models as platforms for identification of novel genes as antimicrobial drug targets," *Future Microbiol.*, vol. 13, no. 4, pp. 455–467, Feb. 2018, doi: 10.2217/fmb-2017-0195.

[33] D.-S. Lee *et al.*, "Comparative genome-scale metabolic reconstruction and flux balance analysis of multiple *Staphylococcus aureus* genomes identify novel antimicrobial drug targets," *J. Bacteriol.*, vol. 191, no. 12, pp. 4015–4024, Jun. 2009, doi: 10.1128/JB.01743-08.

[34] A. M. Abdel-Haleem *et al.*, "Functional interrogation of Plasmodium genus metabolism identifies species- and stage-specific differences in nutrient essentiality and drug targeting," *PLOS Comput. Biol.*, vol. 14, no. 1, p. e1005895, Jan. 2018, doi: 10.1371/journal.pcbi.1005895.

[35] R. A. Rienksma, P. J. Schaap, V. A. P. Martins dos Santos, and M. Suarez-Diez, "Modeling Host-Pathogen Interaction to Elucidate the Metabolic Drug Response of Intracellular *Mycobacterium tuberculosis*," *Front. Cell. Infect. Microbiol.*, vol. 9, May 2019, doi: 10.3389/fcimb.2019.00144.

[36] H. Nouri, H. Fouladiha, H. Moghimi, and S.-A. Marashi, "A reconciliation of genome-scale metabolic network model of Zymomonas mobilis ZM4," *Sci. Rep.*, vol. 10, no. 1, Art. no. 1, May 2020, doi: 10.1038/s41598-020-64721-x.

[37] I. Thiele and B. Ø. Palsson, "A protocol for generating a high-quality genome-scale metabolic reconstruction," *Nat. Protoc.*, vol. 5, no. 1, Art. no. 1, Jan. 2010, doi: 10.1038/nprot.2009.203.

[38] P. D. Karp *et al.*, "Pathway Tools version 23.0 update: software for pathway/genome informatics and systems biology," *Brief. Bioinform.*, vol. 22, no. 1, pp. 109–126, Jan. 2021, doi: 10.1093/bib/bbz104.

[39] L. Heirendt *et al.*, "Creation and analysis of biochemical constraint-based models using the COBRA Toolbox v.3.0," *Nat. Protoc.*, vol. 14, no. 3, Art. no. 3, Mar. 2019, doi: 10.1038/s41596-018-0098-2.

[40] S. M. Keating, B. J. Bornstein, A. Finney, and M. Hucka, "SBMLToolbox: an SBML toolbox for MATLAB users," *Bioinformatics*, vol. 22, no. 10, pp. 1275–1277, May 2006, doi: 10.1093/bioinformatics/btl111.

[41] J. W. Pinney, M. W. Shirley, G. A. McConkey, and D. R. Westhead, "metaSHARK: software for automated metabolic network prediction from DNA sequence and its application to the genomes of *Plasmodium falciparum* and Eimeria tenella," *Nucleic Acids Res.*, vol. 33, no. 4, pp. 1399–1409, 2005, doi: 10.1093/nar/gki285.

[42] M. Kanehisa and S. Goto, "KEGG: Kyoto Encyclopedia of Genes and Genomes," *Nucleic Acids Res.*, vol. 28, no. 1, pp. 27–30, Jan. 2000.

[43] M. S. Skrzypek, J. Binkley, G. Binkley, S. R. Miyasato, M. Simison, and G. Sherlock, "The Candida Genome Database (CGD): incorporation of

Assembly 22, systematic identifiers and visualization of high throughput sequencing data," *Nucleic Acids Res.*, vol. 45, no. D1, pp. D592–D596, Jan. 2017, doi: 10.1093/nar/gkw924.

[44] "The Gene Ontology (GO) database and informatics resource," *Nucleic Acids Res.*, vol. 32, no. Database issue, pp. D258–D261, Jan. 2004, doi: 10.1093/nar/gkh036.

[45] D. S. Wishart *et al.*, "DrugBank: a knowledgebase for drugs, drug actions and drug targets" *Nucleic Acids Res.*, vol. 36, no. Database issue, pp. D901–D906, Jan. 2008, doi: 10.1093/nar/gkm958.

[46] J. C. Xavier, K. R. Patil, and I. Rocha, "Integration of Biomass Formulations of Genome-Scale Metabolic Models with Experimental Data Reveals Universally Essential Cofactors in Prokaryotes," *Metab. Eng.*, vol. 39, pp. 200–208, Jan. 2017, doi: 10.1016/j.ymben.2016.12.002.

[47] A. Kleczkowski, A. Hoyle, and P. McMenemy, "One model to rule them all? Modelling approaches across OneHealth for human, animal and plant epidemics," *Philos. Trans. R. Soc. B Biol. Sci.*, vol. 374, no. 1775, p. 20180255, Jun. 2019, doi: 10.1098/rstb.2018.0255.

[48] N. Jamshidi and A. Raghunathan, "Cell scale host-pathogen modeling: another branch in the evolution of constraint-based methods," *Front. Microbiol.*, vol. 6, 2015, doi: 10.3389/fmicb.2015.01032.

[49] Viana, R., Dias, O., Lagoa, D., Galocha, M., Rocha, I., & Teixeira, M. C. (2020). Genome-Scale Metabolic Model of the Human Pathogen *Candida albicans*: A Promising Platform for Drug Target Prediction. Journal of Fungi, 6 (3), 171. https://doi.org/10.3390/jof6030171.

Chapter 4

# Responses of White Blood Cells to Killed *Candida albicans* as a Preventive Strategy

*Ahmad Ibrahim*

## Abstract

*C. albicans* is by far the most common Candida species causing infection in humans which include superficial and a life- threatening systemic infections. Despite the public health significance of candida infections, phenotypic switching of *C. albicans*, slow mycological diagnosis, limitation of use of antifungal agents due to toxicity, high cost and emergence of resistance have impeded effective treatment. Therefore, a need for safe and potent strategy to prevent this disease is necessary. This chapter discusses the roles of white blood cells as the first line defense mechanism against inactivated *C. albicans*.

**Keywords:** white blood cells, *Candida albicans*, immune response

### 1. Introduction

Fungal infections are a serious public health concerns, particularly with the growing number of immunocompromised individuals. *C. albicans* amongst other fungal species has been identified as one of the leading cause of infections in recent times [1]. Candidemia and Candidiasis account for 50% and 70% prevalence infections in human and has caused a great deal of morbidity and mortality largely because of the polymorphic nature of *C. albicans*. Also, factors such as toxicity of antifungal drugs, drug resistance, limited arsenal of antifungal drugs, slow mycological diagnosis, variable drug bioavailability in immune-compromised patients and drug interactions have truncated every efforts being made at mitigating the prevalence and its consequential effects. These challenges have led to the several attempts at developing a viable preventive option for candida infections. The cellular surface of *C. albicans* is a predominant source of immuno-stimulatory antigens [2] comprising of 90% carbohydrates and 10% proteins [3] and hence making carbohydrates dominate immune recognition while the proteins exhibit the key role of adhesive interactions with the host cellular surfaces. Therefore, complete inactivation of the pathogen amongst many strategies such as using the genetic material, a specific protein on the cell surface or attenuation to prevent candidiasis has been attempted. The killed *C. albicans* has lost its ability to infect their hosts but can stimulate enough immunological responses for the host protection. White blood cells response against *C. albicans* is initiated within the first few hours of inoculation or infection.

Therefore during host – killed *C. albicans* interaction, the cell surface molecules trigger and modulate cell mediated (T cells) and innate immune cells (macrophages, neutrophils and natural killer cells) to respond appropriately. These cells are considered to be the first line and most important defense mechanism against Candidiasis [4] and consequently induced a strong response against the pathogen [5]. These responses function synergistically, co-operate and modulate each other with the final goal of fighting infection (4).

## 2. Recognition of *C. albicans* in mucosal surfaces

The epithelial cells represent the first line of defense against Candida infection on mucosal surfaces. As the predominant cells in the innate immunity of the host, epithelial cells express pattern recognition receptors, which recognize *C. albicans* by interacting with pathogen-associated molecular patterns on the fungal cells. However, there are three major groups of these receptors (Toll-like receptors, C-type lectin receptors and nod-like receptors) but only certain Toll-like receptors and C-type lectin receptors on epithelial surfaces recognize *C. albicans*. In addition to pattern recognition receptors, other cell-surface proteins, such as E-cadherin and Epidermal Growth Factor Receptor, can also recognize Candida and these are unsurprisingly implicated in Candida adherence and endocytosis [6, 7].

Candida albicans expresses proteins called adhesins for attachment with specific receptors on epithelial mucosa depending morphology

A. Adhesion

Invasin proteins mediate penetration of C. albicans into the host cell

B. Invasion

Recognition of pathogen by different patterns recognition receptors (PRRs) of white blood cells specifically the monocytes or macrophages, neutrophils and dendritic cells

C. Recognition and phagocytosis

Cytokines production and presentation of antigen by macrophages

D. Cytokinesis

Acquired immune response mediated by Th1 and Th2 activate B cells leading to killing of C. albicans

E. Stimulation of adaptive immunity

**Figure 1.**
*Summary of host immune response against* C. albicans.

### 2.1 Host immune response

The cell surface receptors on *C. albicans* initiate adhesive interactions and invade the host cell using a series of proteins including adhesins and invasins. These immunodominant factors would trigger and stimulate a complex interplay of natural and adaptive immunity, posing interesting immunological response to the host. Cell-mediated (T cells) and innate immunity (macrophages, neutrophils and natural killer cells) are considered to be the most important line of defense against candidiasis [4] as they are recruited into the site of infection to exert protective effects. These include phagocytosis and antigen presentation, opsonization and production of chemicals for effective killing of the microbial cells. It must be emphasized therefore, that these responses comprise of different arms of the immune system (innate, cell-mediated and antibody-mediated) as shown in **Figure 1**.

#### *2.1.1 Innate immunity*

White blood cells are produced and derived from multi-potent cells in the bone marrow known as hematopoietic stem cells and are found throughout the body, including the blood and lymphatic system [8]. Five individual types of white blood cells namely neutrophil, monocytes, lymphocytes, basophils, eosinophils [9] are involved in sustaining immunity [10].

Innate immune response is the dominant protective mechanism against disseminated candidiasis [11] and host defense against fungal infection depends on elimination of the fungi by phagocytic cells of the innate immune system, especially neutrophils and macrophages [12] at the initiation of infection before other immune cells are mobilized. Therefore, white blood cells are used to assess the working condition of body's immune system, to determine an active or chronic infection, identify the type of infection and also point to an allergic response or inflammation in the body [9].

Hence, quantitative and qualitative abnormalities of these immune cells are indications to different physiological conditions and particularly neutrophils and monocytes are associated with systemic candidiasis.

##### *2.1.1.1 Neutrophils*

Neutrophils or Polymorphonuclear leukocytes are the predominant phagocytic immune cells that play a major role against *C. albicans* infection. These cells activate various antimicrobial mechanisms in addition to phagocytosis, such as producing reactive oxygen species, the release of granular enzymes and antimicrobial proteins [13]. In addition, a neutrophil extracellular trap composed of a neutrophil chromatin is another significant protective strategy deployed by the host against fungal infections [14]. Neutrophils and monocytes damage and kill yeast cells of *C. albicans*, hyphae and pseudohyphae [11] by recognizing and engulfing opsonized and non-opsonized yeast cells via cell-surface pattern recognition receptors. However, the large size of Candida hyphae and pseudohyphae may preclude phagocytosis and thus the need for several phagocytes to collaborate and affect extracellular killing [15].

##### *2.1.1.2 Monocytes*

Neither dead cell debris nor attacking microorganisms can be dealt with effectively by the neutrophils [16]. Monocytes and their derivatives, including macrophages and dendritic cells, play diverse roles in the response to fungal pathogens by sensing fungi and triggering signaling pathways that mediate direct effects like phagocytosis, cytokine production and presentation of fungal antigens to elicit adaptive immune response [17].

- **Phagocytosis**

  In phagocytosis, fungi can be eliminated in monocytes and their derivatives in the phagolysosome. This is an acidified compartment that contains enzymes such as Nicotinamide Adenine Dinucleotide Phosphate (NADPH) oxidase (generates reactive oxygen species), and inducible nitric oxide synthase (produce Nitrogen IV Oxide and reactive nitrogen species) that can sequester nutrients and in response to pro-inflammatory stimuli [18]. This fungal killing may be sufficient to halt the progression of infection, but it can also provide fungal antigens that can be used to initiate the adaptive immune response to ensure sterilizing immunity. Fungal uptake is not always beneficial to the host, however, as some fungi have adapted to the harsh environment in the phagolysosome or can subvert monocytes to enable fungal persistence and proliferation [17].

- **Production of Cytokines**

  Cytokines are a group of low molecular weight proteins that act as a mediator between cells and are produced by white blood cells and other non-immune cells in response to stimuli [19].

  Monocytes and their derivative cells can produce chemokines, pro-inflammatory, anti-inflammatory and pleiotropic cytokines [20, 21]. These cytokines and chemokine secretion is important for the development of both the innate and adaptive immune response to fungal pathogens and can influence the activation and recruitment of other immune cells and the polarization of the adaptive immune response [17]. Under normal physiological condition, cytokines are not detectable or are present at low levels in body fluids or tissues because they are only produced when required in immune responses [19]. Therefore, an elevated levels or unregulated production of cytokines may be associated with inflammation or disease pathogenesis [22].

- **Presentation of Antigens**

  Antigen-Presenting Cells (APCs) are cells that can process a protein antigen, break it into peptides, and present it in conjunction with class II Major Histocompatibility Complex (MHC) molecules on the cell surface where it may interact with appropriate T cell receptors. Monocytes and their derivative are professional APCs and are amongst the principal antigen-presenting cells for T cells [23]. Antigen-Presenting Cells are critical for the initiation of adaptive immune responses and for maintenance of peripheral tolerance [24]. Dendritic cells serve as the connection between innate and acquired immunity and morphological characteristics of *C. albicans* dictates the specific immune response [25]. For example, the interaction of dendritic cells with yeast cells or pseudohyphal sensitize different receptors. Therefore, when yeast form of *C. albicans* is engulf by dendritic cells, differentiation of CD4+ cells into T-helper 1 cells is induced, while dendritic cells stimulated by the pseudohyphal form induce a T- helper 2 response. The response produced by

T- helper 1 cells is linked with protection of the host against fungal infection while for T- helper 2, responses are related to the ability of microorganisms to escape or suppress the host's immune response. Nonetheless, T-helper 1 and T-helper 2 responses activate B cells and leads to maturation of other phagocytic cells [23].

#### *2.1.1.3 Eosinophils and basophils*

Eosinophils and Basophils are both granulocytes characterized by their content of intracellular granules. These cells become especially active during an allergic response and are responsible for releasing histamine [9]. Fungi also represent a source of major allergens.

While the roles of eosinophils in an allergic disease associated with fungal sensitization is still debatable or even poorly understood, their contributions to remodeling are more accepted.

### 2.2 Inactivation of *Candida albicans*

The commonest methods for inactivation of *C. albicans*, in the preparation of an immune-based prevention of *C. albicans* infection include using of heat and a source of UV. According to Evron, [26], whole *C. albican* cells suspended in 0.85% sterile normal saline and heated at 65°C are inactive. While exposing *C. albicans* suspended in 0.85% sterile normal saline directly to a source of Ultraviolet radiation (UV) at a wavelength of 254 nm for 30 minutes inactivates the cells [27]. This inactivation renders the *C. albicans* non-viable to infect an intended host but retains the structural conformation of immunogenic components on the cell surface.

#### *2.2.1 Response of antibodies to Killed* C. albicans

Adherence of lymphocytes to a fungus is the first step in the direct lymphocyte-mediated anti- fungal effect against *C. albicans* [1]. Experimental study indicates that antibodies play an important role in host defense against disseminated candidiasis because individuals with defects in cell mediated immunity mechanisms are particularly prone to superficial but not disseminated candidiasis [28].

Therefore, humoral mediated immune response results in a significant elevated level of antibodies in Wister albino rats exposed to killed *C. albicans*. This could be as a result of recognition of the immunogenic proteins and glycoproteins on the cell surface and subsequent stimulation of memory cells to produce significant quantity of antibodies on a second encounter of similar antigens on killed *C. albicans* that are immunoprotective. According to Evron, [29], circulating antibodies in mice exposed to killed - *C. albicans* that are immunoprotective should be greater than 256 μg/m. Hence a concerted effort for more research to produce vaccines that can stimulate the release of even more antibodies in rats and subsequently in human are necessary.

#### *2.2.2 Response of phagocytic cells (monocytes, macrophages and granulocytes)*

Phagocytic cells such as the granulocytes and monocytes play an important role in cell-mediated immunity (T-cells and phagocytic cells such as monocytes, granulocytes) and so attacks the killed *C. albicans* in similar mechanism as though it is viable and infectious cells. These cells are the first line defense mechanism and are recruited in large quantity in the first few days of injection of the killed *C. albicans* to the Wister rats. Granulocytes being components of white blood cells are recruited as innate immune response to engulf the killed *C. albicans* [29]. However, viable *C. albicans* have the ability of switching or morphogenesis which allows them to escape phagocytosis by piercing of phagocytes and subsequent killing of phagocytic cells, leading to a decrease in circulating granulocytes in blood [30]. Consequently, contributing to a large extent the impeding factor for availability of an effective vaccine.

*2.2.3 Delayed-type hypersensitivity*

Delayed-Type Hypersensitivity reaction is initiated when antigens are presented by antigen presenting cells (i.e. langerhans cells) to sensitized memory T cells. The antigen presentation and subsequent T cell activation elicit an influx of macrophages, monocytes and lymphocytes at the site of antigen exposure. At the onset of delayed-type hypersensitivity reaction, verso-permeability is increased so that additional cellular components migrate into the local site of antigen presentation [31] and this explains the swelling at the site of injection of killed *C. albicans*. Therefore, inactivated *C. albicans* have an immune-stimulatory property.

## 3. Conclusion

The need for appropriate immuno-prophylaxis or immunotherapy against candidiasis is readily apparent. Therefore, the relationship between killed *Candida albicans* and the hosts' white blood cells in terms of recognition and response clearly suggest an interesting immunoprotection against viable *C. albicans.*

A safe and effective therapeutic alternatives to combat these infections and to eliminate potential problems of toxicity and emergence of resistance to the limited options of antifungal drugs is needed [32]. Therefore, killed *C. albicans* is an immune-based prophylactic and therapeutic approach [33] which represents novel option against *C. albicans* infections [32].

**Thanks**

I want to specially thank my brother, Muhammad I. Odaki for his tremendous support.

## Author details

Ahmad Ibrahim
Department of Biochemistry, Federal University Lokoja - Kogi State, Nigeria

*Address all correspondence to: ahmadibrahim337@yahoo.com

## References

[1] Forsyth C. B. and Mathews, H. C. (2002). Lymphocytes adhesion to C. albicans. 70(2): 517-527.

[2] Martınez, J.P., Gil, M.L., Lopez-Ribot, J.L. and Chaffin, W.L. (1998) Serologic response to cell wall mannoproteins and proteins of Candida albicans. Clin. Microbiol. Rev. 11, 121-141.

[3] Staib P, Morschhäuser J. (2007). "Chlamydospore formation in Candida albicans and *Candida dubliniensis* an enigmatic developmental programme". Mycoses 50 (1): 1-12. PMID 17302741.

[4] Levitz, S.M. (1992). Overview of host defenses in fungal infections. Clin. Infect. Dis. 14 (Suppl 1), S37–S42.

[5] Casadevall, A., Cassone, A., Bistoni, F., Cutler, J.E., Magliani, W., Murphy, J.W., et al., (1998). Antibody and/or cell-mediated immunity, protective mechanisms in fungal disease: an ongoing dilemma or an unnecessary dispute? Med. Mycol. 36, 95-105.

[6] Phan QT, Myers CL, Fu Y, Sheppard DC, Yeaman MR, Welch WH, et al., (2007). Als3 is a Candida albicans invasin that binds to cadherins and induces endocytosis by host cells. PLoS Biol. 5:e64.

[7] Sun JN, Solis NV, Phan QT, Bajwa JS, Kashleva H, Thompson A, Liu Y, Dongari-Bagtzoglou A, Edgerton M, Filler SG (2010). Host cell invasion and virulence mediated by Candida albicans Ssa1.PLoS Pathog. 6:e1001181.

[8] Maton, D., Hopkins, J., McLaughlin, Ch. W., Johnson, S., Warner,et al.,(1997). Human Biology and Health. Englewood Cliffs, New Jersey, US: Prentice Hall. ISBN 0-13-981176-1

[9] D'Aquila R., (2011) – How to Interpret Your Blood Tests: Part II. NYC Chiropractor – Applied Kinesiology. https://robdaquila.com/2011/03/15/how-to-interpret-your-blood-tests-part ii/received 15/03/2011.

[10] Diong, K., and Thompson, L. A., (2017). A methodical approach to interpreting the white blood cell parameters of the complete blood count. American Society of Clinical Laboratory Science. 30(3): 186-193

[11] Diamond, R.D., Clark, R.A. & Haudenschild, C.C. (1980). Damage to Candida albicans hyphae and pseudohyphae by the myeloperoxidase system and oxidative products of neutrophil metabolism in vitro. Journal of Clinical Investigation, 66, 908-917.

[12] Blanco JL, Garcia ME (2008). Immune response to fungal infections. Vet Immunol Immunopathol. ;125:47-70.

[13] Robinson JM (2008). Reactive oxygen species in phagocytic leukocytes. Histochem Cell Biol. 130:281-297.

[14] Urban CF, Reichard U, Brinkmann V, Zychlinsky A., (2006). Neutrophil extracellular traps capture and kill Candida albicans yeast and hyphal forms. Cell Microbiol. 8:668-676.

[15] Bellocchio, S., Montagnoli, C., Bozza, S., Gaziano, R., Rossi, G., Mambula, et al., (2004). The contribution of the Toll-like/IL-1 receptor superfamily to innate and adaptive immunity to fungal pathogens in vivo. Journal of Immunology, 172, 3059-3069.

[16] Sozzani, S.; Zhou, D.; Locati, M.; Bernasconi, S.; Luini, W.; Mantovani, A.; O'Flaherty, J. T. (1996). "Stimulating properties of 5-oxo-eicosanoids for human monocytes: synergism with monocyte chemotactic protein-1 and -3".

The Journal of Immunology. **157** (10): 4664-4671

[17] Heung, L.J. (2020). Monocytes and the host response to fungal pathogens. Front. Cell infect. Microbial. 10:34

[18] Uribe-Querol E., Rosales C. (2017). Control of phagocytosis by microbial pathogens. Front. Immunol. 8:1368. 10.3389/fimmu.2017.01368

[19] Chin, VK., Foong, KJ. Maha, A. Rusliza, B. Norhafizah, M. Chong, P.P (2014). Early expression of local cytokines during systemic *Candida albicans* infection in a murine intravenous challenge model, Pages: 869-874

[20] Carson W. E., Ross M. E., Baiocchi R. A., Marien M. J., Boiani N., Grabstein K., et al. (1995). Endogenous production of interleukin 15 by activated human monocytes is critical for optimal production of interferon-gamma by natural killer cells *in vitro*. J. Clin. Invest. 96, 2578-2582.

[21] Arango Duque G., Descoteaux A. (2014). Macrophage cytokines: involvement in immunity and infectious diseases. Front. Immunol. 5:491.

[22] Abbas AK, Lichtman AH and Pober JS: Cellular and Molecular Immunology. 3rd edition. W.B. Saunders; Philadelphia, PA: pp. 15-37. 1997

[23] Cruse, J. M., Lewis, R. E. & Wang, H. (2004). Antigen Presentation: Immunology Guide Book. Academic Press. 267-276

[24] Dana, R. & Hamrah, P., (2010). APCs in the Eye and Ocular Surface. Encyclopedia of the Eye. 120-127.

[25] Ranta, K. Nieminen, T. Saariaho et al., (2013). "Evaluation of fungal extracts to determine immunomodulatory properties," Journal of Investigational Allergology and Clinical Immunology, vol. 23, no. 4, pp. 226-233.

[26] Evron Ruth (1980). In Vitro Phagocytosis of *Candida albicans* by Peritoneal Mouse Macrophages. *infection and immunity*, p. 963-971 Vol. 28, No. 3. 0019-9567/80/06-0963/09$02.00/0

[27] Wheeler, RT., and Fink, G.R. (2006). A drug –sensitive genetic network masks fungi from the immune system. PLOS Pathogens, 0328-0338

[28] Maathew R and Burnie, (2001). Antifungal antibodies: a new approach to the treatment of systemic candidiasis. Current opinion in investigational drugs 2:472-476

[29] Aderem A. (2003). Phagocytosis and the Inflammatory Response. The Journal of Infectious Diseases, Volume 187, Issue Supplement_2, 15 June 2003, Pages S340-5, https://doi.org/10.1086/374747

[30] Uwamahoro, N., Verma-Gaur, J., Shen, H. H., Qu, Y., Lewis, R., Lu, J., Bambery K., *et al*., (2014). The pathogen C. albicans hijacks pyroptosis for escape from macrophages. American Society of Microbiology. 5(2): 1-11

[31] Jyonouchi Harumi, (2015). Delayed Type Hypersensitivity. Emedicine. medscape.com/article/886393-overview.

[32] Wang *X., Sui, X. L. E., & Jiang Y. (2015). Vaccinces in the treatment of invassive land disease.* Virulence *6(4):309-315*

[33] Deepe Jr., G.S. (1997). "Prospects for the development of fungal vaccines". Clin. Microbiol. Rev. 10:585-596.

Section 3

# Pathogenicity of *Candida albicans*

Chapter 5

# Candida Onychomycosis: Mini Review

*Sandra Widaty, Eliza Miranda and Caroline Oktarina*

## Abstract

Onychomycosis is a common fungal infection affecting nails. The infection is frequently due to dermatophyte, while yeast and non-dermatophyte molds (NDMs) attributed especially in immunocompromised patients. NDMs and *Candida* species can be involved as primary or secondary pathogens. Candida onychomycosis (CO), most commonly caused by *C. albicans* and *C. parapsilosis*, is frequently associated with local or systemic immune disturbances. In the cases that the host immunity is severely affected, *Candida* acts as primary pathogen, while other diseases e.g., diabetes mellitus, malnutrition, and smoking serve as predisposing factors for Candida to cause secondary infection. Furthermore, formation of biofilms and production of enzymes contribute as the virulence factors of the yeasts. Clinical manifestation of CO varies, from discoloration and marked thickening of the nail to dystrophic nails with fingernails more commonly affected. Paronychia is the most common type of CO and Candida granuloma is one of the severe types of CO which often occurs in chronic mucocutaneous candidiasis. Establishing the diagnosis of CO is crucial as well as the identification of each predisposing factors. Microscopic examination and fungal cultures are the gold standard examination for diagnosing onychomycosis, while for NDM, multiple confirmation and repeated examination is needed due to its as contaminants.

**Keywords:** candida onychomycosis, nail, fungal infection, diagnostic challenges, treatment

## 1. Introduction

Onychomycosis is a common nail infection caused by fungi, namely yeasts, dermatophytes, and non-dermatophyte molds (NDMs) [1]. The prevalence keeps increasing with age and it is commonly identified in elderly populations. Approximately 20% of adults in their second to fourth decades are affected by this disease. Yeasts contribute to 24–50% cases of onychomycosis with *Candida* species as the most common agent [2]. Various factors are associated with the event of onychomycosis, e.g. host's comorbidities (human immunodeficiency virus [HIV] infection, diabetes mellitus, peripheral circulation disturbances), repeated nail trauma, smoking, antibiotic therapy, immunosuppressive therapy, repeated exposure to fungi, humid climates, genetic predisposition, and occlusive footwear [2, 3]. Establishing the diagnosis of Candida onychomycosis (CO) is challenging. Albeit frequently identified in culture and direct microscopic examination, the presence of *Candida* species might only be colonization, not necessarily the cause of nail diseases. A careful interpretation of diagnostic tests' results and correlation with results of clinical examination are necessary to establish the diagnosis of CO which will aid the clinicians in providing correct treatment for the patients [4].

## 2. Epidemiology

The prevalence of onychomycosis differs based on geographical location with worldwide prevalence of approximately 10% [5]. The incidence onychomycosis in North America ranges from 8.7–13.8% while the prevalence in Southeast Asia ranges from 2–6% [2, 3]. Higher prevalence is reported in countries with humid climates, such as Greece (27.99%) and Ethiopia (60,4%), [6, 7]. While there are three groups of fungi responsible for onychomycosis, dermatophytes are the most common cause of onychomycosis (60–70%) [3].

Yeasts, most commonly identified as *Candida* species in onychomycosis, contributes in approximately 40% of the onychomycosis cases in Southeast Asia. Other studies reported that the prevalence of CO varies between 24–50% of onychomycosis cases [2]. NDMs and yeasts onychomycosis is more common in subtropical and tropical climates while dermatophytes is more common in temperate climates [7]. *Candida albicans* is identified as the most common isolated species, followed by *C. parapsilosis, C. krusei, C. tropicalis,* and *C. glabrata* [2]. CO is more frequently to be identified in fingernails than toenails, especially in patients with hands continuously immersed in water [3].

## 3. Prognostic factors

In assessing the treatment outcome of CO patients, there are three types of cure to be considered, which area mycological cure, clinical cure, and complete cure. Clinical cure is described as a previously infected nail without signs and symptoms of onychomycosis. Mycological cure is described as negative results on both direct microscopic examination and culture. Complete cure is described as achieving both clinical and mycological cure [8]. Various prognostic factors have been identified for the treatment outcome of onychomycosis. In general, they can be divided into three groups, which are patient's characteristics, nail features, and the infectious agents (**Table 1**) [9].

Most studies reported that the onychomycosis is more commonly diagnosed in men. Male patients are associated with poor outcome because they are more likely to be exposed to repeated nail trauma and they usually do not seek health care until the disease becomes too advanced. Furthermore, they are more likely to have low compliance; hence, male patients become more resilient when it comes to treatment and have 2,6 times risk of not achieving clinical cure [8]. Increasing age is also known to be associated with poor prognosis in onychomycosis patients because elderly populations usually suffer from poor circulation system, poor immune status, decreased nail growth, and mixed fungal infections. Hence, their response to therapy might be lacking and they have 3,7 times risk of not achieving clinical cure [8, 9].

Nail trauma can exert significant and irreversible damage which will predispose patients to onychomycosis. Patients who have abnormal nails with positive mycology examination had 5,4 times risk of developing onychomycosis [9]. Other poor prognostic factor is history of onychomycosis. Patients with prior infection have 2,3 times risk of not achieving clinical cure. These patients are more likely not to respond standard treatment course since they have been treated before. There might also be involvement of genetic susceptibility in the development of recurrent onychomycosis [8].

| **Poor Prognostic Factors** | |
|---|---|
| Patient's characteristics | Male gender |
| | Increasing age |
| | History of nail trauma |
| | History of onychomycosis |
| | Poor immune status |
| | Poor peripheral circulation |
| | Uncontrolled diabetes mellitus |
| | Repeated exposure to water and detergents |
| | Repeated exposure to mud and soil |
| | Barefoot walking |
| Nail features | Subungual hyperkeratosis >2 mm |
| | Fingernail and toenail involvement |
| | More than 3 infected nails |
| | Matrix involvement |
| | Significant lateral disease |
| | Dermatophytoma |
| | Nail plate involvement >50% |
| | Slowly growing nails |
| | Hallux involvement |
| | Severe onycholysis |
| | Paronychia |
| | Melanonychia |
| | Total dystrophic onychomycosis |
| Infectious agents | Fungal and bacterial coinfections |
| | Yeasts |
| | Non-dermatophytes molds |

**Table 1.**
*Poor prognostic factors in onychomycosis [4, 8–10].*

As most fungi are opportunistic agents, poor immune status can predispose patients to onychomycosis, especially in HIV patients with CD4 count <400/mm$^3$. The onychomycosis is more likely to involve fingernails and toenails also more severe [9]. Patients with hand and foot involvement have 1.1 times risk of not achieving complete cure and if the patients have more than 3 infected nails, they have 1.5 times risk of not achieving complete cure [4]. Furthermore, hallux involvement presents as poor prognostic sign because it is more likely to suffer repeated trauma lead to predisposition of continuous infection.

In addition, poor peripheral circulation caused by chronic venous disease is associated with poor prognosis. Chronic venous disease can cause nail dystrophy, hyperkeratosis, discoloration, hyperplastic nail bed, and onychogryphosis. Only 25% of patients treated with itraconazole are cured [9]. Poor peripheral circulation can also identified in patients with uncontrolled diabetes mellitus which associates with secondary infections and nonhealing ulcers. This population is also reported to have more severe onychomycosis, high recurrence rate and longer duration to achieve complete cure [9].

Repeated exposure to water and detergents will predispose patients to chronic paronychia and affect the drug delivery since the tissue is more edematous and inflamed. While repeated exposure to mud and soil, also barefoot walking will predispose the patients to repeated minor trauma. Most fungi are saprophytic; hence, they can invade nails easily in this condition. This often happened in tropical countries [10].

Subungual hyperkeratosis is host's reaction towards fungal infection by thickening the stratum corneum. The thick debris presents as a barrier to antifungal agents, both systemic and topical agents [9]. Furthermore, patients with matrix involvement have 2.1 times risk of not achieving complete cure [8]. Matrix is known to be the nail's origin [11]. Hence, matrix involvement in onychomycosis will affect the nail growth and drug delivery [8]. In addition, the slow nail growth is also a poor prognostic factor since the patients shed the infected portion of the nail more slowly. This association is also described in elderly populations. Slow nail growth is also seen in significant nail plate involvement. Greater surface involved is associated with greater fungi load; hence, lower cure rates [9].

Significant lateral disease affects the treatment outcome since there is poor attachment of the lateral edge to the nail groove. This can reduce the drug delivery about two thirds of normal nail. Similar cases are seen in severe onycholysis [9]. Patients with lateral disease have 3,5 times risk of not achieving complete cure [8]. Another poor prognostic factor is dermatophytoma, a dense thick-walled fungal elements presenting as white to yellow patch or longitudinal streak in nail plate. This dense mass is difficult to be penetrated by antifungal agents. Therefore, the patients with dermatophytoma have 2.9 times risk of not achieving clinical cure [8, 9].

Melanonychia is black pigmentation identified on the nail plate. This feature is associated with poor prognosis in onychomycosis. However, the association has not been elucidated yet. Total dystrophic onychomycosis (TDO) is the final destructive stage of onychomycosis, in which there is thickened nail bed, crumbling nail plate, and significant involvement of nail matrix. Patients with TDO have 1.1 times risk of not achieving complete cure [4, 9].

As for the infectious agents, CO and NDMs onychomycosis indicate poor prognosis. CO is associated with immunosuppression, especially in case of chronic mucocutaneous candidiasis (CMC) and HIV patients. While NDMs infections are difficult to be diagnosed and lack of data for treatment course. In addition, fungal and bacterial infections can complicate the treatment. Therefore, these factors can implicate in poor prognosis of onychomycosis patients [9].

In order to aid the clinicians in have better treatment outcome, several instruments have been developed to predict the prognosis in onychomycosis patients. The first developed instrument was Scoring Clinical Index for Onychomycosis (SCIO Index). This scoring assesses the nail's clinical component based on its clinical form, nail involvement, and subungual hyperkeratosis. In addition, it assesses the growth component based on the patient's age and location of onychomycosis. As the score increases, the onychomycosis might be more difficult to treat [12]. However, this scoring has not been validated and has other limitations, such as exclusion of important prognostic factors and complex calculation [9].

Another scoring was developed by Baran et al. (**Table 2**). The higher the score, the more likely the treatment failure will happen [13]. Albeit being the most comprehensive instrument, this index has not been validated and time-consuming [9].

The most commonly used instrument is Onychomycosis Severity Index (OSI). OSI is simpler by assessing three major components, which are area of involvement, proximity of disease to matrix, presence of dermatophytoma or subungual

| | Descriptor | Subdivision | Score |
|---|---|---|---|
| 1 | Extent of involvement | Distal one-third of nail plate | 1 |
| | | Distal two-thirds of nail plate | 2 |
| | | Proximal nail plate involvement | 3 |
| 2 | Diffuse nail plate thickening | Mild or moderate | 1 |
| | | Associate with onychogryphosis | 3 |
| 3 | Nail plate thickening associated with the appearance of linear streaks – includes the change confined to the lateral border | One streak only | 2 |
| | | Two or more streaks | 3 |
| | | If the streaks are black do not score but see 7 | |
| 4 | Onycholysis | Affecting the distal two-thirds of nail plate | 2 |
| 5 | Location | Any one of: | |
| | | Second to fifth toes or thumb | 1 |
| | | Great toenail | 2 |
| 6 | Paronychia associated with nail plate disease | With diffuse melanonychia | 3 |
| | | With melanonychia at the lateral edges of the nail | 3 |
| 7 | Melanonychia (without paronychia) | With one or more longitudinal streaks | 3 |
| | | Diffuse pigmentation | 4 |
| 8 | Age of patient | Under 7 years | 3 |
| | | 7–25 years | 1 |
| | | 25–60 years | 2 |
| | | Over 60 years | 3 |
| 9 | Presence of the following predisposing factors | Diabetes mellitus | 1 |
| | | Known severe trauma to affected nail | 2 |
| | | Immunosuppression (due to therapy, e.g., prednisolone, or disease, e.g., AIDS) | 4 |
| | | Symptomatic peripheral vascular disease | 2 |
| 10 | Causative organism | *Scytalidium* spp. | 4 |
| | | Other mold fungi | 2 |
| | | Yeasts | 1 |

*Cited as is from Baran et al. [13].*

**Table 2.**
*Baran-Hay's severity index for onychomycosis [13].*

hyperkeratosis >2 mm (**Table 3**). The score is multiplication of score for area of involvement with score for proximity of disease and addition of score for the presence of dermatophytoma or subungual hyperkeratosis >2 mm. Score 1–5 indicates mild onychomycosis; 6–15 indicates moderate onychomycosis; and 16–35 indicates severe onychomycosis [14]. This index has been validated with high reliability. However, this index only assesses one nail, does not correlate the severity of disease with treatment outcome, and excludes other important prognostic factor [9].

| Predictor | Subdivision | Score |
|---|---|---|
| Area of involvement (%) | 0 | 0 |
| | 1–10 | 1 |
| | 11–25 | 2 |
| | 26–50 | 3 |
| | 51–75 | 4 |
| | 76–100 | 5 |
| Proximity of disease to matrix | <1/4 | 1 |
| | 1/4–1/2 | 2 |
| | >1/2–3/4 | 3 |
| | >3/4 | 4 |
| | Matrix involvement | 5 |
| Presence of dermatophytoma or subungual hyperkeratosis >2 mm | No | 0 |
| | Yes | 10 |

*Cited as is from Carney et al. [14].*

**Table 3.**
*Onychomycosis severity index [14].*

## 4. Pathogenesis and causative agent

The most common causative agent of yeast onychomycosis is candida species. Fingernails are the predilection site of CO, especially in patients who are regularly submerging their hands in water [3]. Candida species are a commensal part of the normal skin flora, which are present in nature. However, these species may exhibit opportunistic quality in an immunocompromised host. Candida species can be either primary or secondary causative agent in onychomycosis. Primary CO can be commonly encountered in a severe immunocompromised host, for example, in HIV patient. On the contrary, secondary CO is usually related to predisposing diseases or circumstances, for instance, diabetes mellitus, malnutrition, peripheral vascular disease, chronic nail trauma, smoking, and vulnerable age (elderly and children). Particular occupations such as housekeepers, fishers, and farmers are also at risk of CO due to the frequent trauma and excessive moisture on the nails, exposure to contaminants, and contact with chemicals [2].

Instead of appearing as individual spores and hyphae, fungal organisms tend to integrate, forming a biofilm which is a syntrophic group of fungi adhering to the host's surface. When not infiltrating a substrate, fungi may fluctuate between free-floating types and parts of a superficial biofilm. This particular feature provides benefits for fungi development while being surrounded by extracellular matrix (ECM). The surrounding ECM defends fungi from the host's immune response and antifungal treatments. ECM also supports fungi to distribute nutrients to the biofilm. Fungi biofilm contributes to the rationale of why onychomycosis is relatively refractory to antifungal treatment and challenging to eliminate the spores in chronic manifestation entirely [3].

The biofilm development by *C. albicans* is initially started with the adhesion and colonization of *C. albicans* cells on an appropriate substrate. Several features that influence the attachment process of *C. albicans* are non-specific factors (electrostatic forces and hydrophobic part of the cell membrane) and specific factors (adhesin on the extracellular layer of *C. albicans,* which connects the ligands on the film). Besides attaching to their counterparts, *Candida* species can also occur secondary to bacteria that have previously colonized their host. When the attachment process followed by microcolonies formation has completed, *C. albicans* began to proliferate characterized by the budding yeasts, production of filamentous structure, and deposition of ECM materials, ultimately resulting in biofilm formation. The filamentous structure supports the biofilm scaffolding and protects the adhesion spots for the budding yeasts [15].

There are 3 definite stages of *C. albicans* observed through microscopic examination, including early stage (0–11 hours), intermediate stage (12–30 hours), and maturation stage (72 hours) [15, 16]. During the 3rd and 4th hour, budding yeasts' microcolonies can be observed, while pseudo-hyphae and true hyphae start to appear in the 4th hour and 8th hour, respectively. Throughout the intermediate stage, microcolonies are later bounded by hyphae, which eventually results a single coalescent layer formation. An opaque film overlaying the microcolonies can be observed at this stage, which is mainly composed of non-cellular material such as polysaccharides. The basal layer is composed of yeast cells, while the filamentous cells constitute the underlying structure. Eventually, the maturation stage is characterized by the multiplication of extracellular material in a time-dependent manner until the mature biofilm covering the entire fungi has been developed [15].

In addition to biofilm, yeast factors that contribute to the virulency of CO are synthetization of hydrolytic enzymes, including proteinase, hemolysin, and phospholipase, which are unique between each type of Candida species. Moreover, proteinase plays a part in the breakdown of protein and phospholipase contributes to the destruction of the host cell, allowing *Candida* species to invade the host [2]. The reported prevalence of CO is 5–10% of all onychomycosis cases. The most common causative species of CO are *C. albicans* and *C. parapsilosis* [17].

## 5. Clinical presentation in immunocompetent and immunocompromised patients

Most CO cases involve fingernails compared to toenails, with an estimated prevalence of up to 50% of onychomycosis cases in fingernails. Women at risk of developing CO are typically wet workers due to the recurrent moist in the hands, exposure to trauma, regular contact with washing liquids, and contamination to vaginal flora during cleansing, which ultimately provides a suitable niche for the development of Candida species [2].

Clinical presentations that are predictive for CO are nail plate dystrophy and off-white discoloration, commonly followed by pigmentation. Melanisation, one of the virulence factors for Candida, suggests an indication of progressive resistance to antifungal treatment. Classification of CO is established because of the complex etiopathogenesis and diverse clinical presentations. The first clinical classification of CO was suggested based on the clinical presentation, the affected location, and the infection route, which are Candida paronychia, Candida granuloma, and Candida onycholysis [2].

The most frequent type of CO is paronychia. Humidity plays an essential role in the development of Candida paronychia. Clinical manifestation of Candida paronychia comprises erythema and swelling in the nail folds followed by gradual dystrophy in the nail plate accompanied by paronychia and Beau lines, which is depicted by an oblique dent in the nail plate suggesting parasite infestation on the nail matrix. The most severe type of CO is granuloma, which is frequently observed in patients with chronic mucocutaneous candidosis. Clinical manifestation of Candida granuloma displays brittle nails and a deformity resembling drumstick which is also referred to as pseudoclubbing. The last type of CO is onycholysis. Clinical manifestation of Candida onycholysis is characterized by subungual distal hyperkeratosis, which further develops into a group of keratosis separating the nail plate from the bed. Moreover, a recent classification was proposed, including four clinical groups of CO, which are chronic paronychia with secondary nail dystrophy, distal onychomycosis, chronic mucocutaneous candidosis, and secondary candidosis [17].

Chronic paronychia initially emerges from the proximal nail fold, although lateral nail folds are occasionally affected in the beginning. Swelling of the periungual skin and a noticeable gap between the fold and nail plate is observed, followed by the nail plate involvement. Marks with a white, green, or black color can be detected at the lateral and distal parts, respectively. The longitudinal ridges and opaqueness appear on the nail that develops into a brittle and easily detached nail. Pressure and movement on the nail can be painful in contrast to dermatophyte infections. A superimposed infection caused by bacteria into the subcuticular space usually occurs, leading to a vicious cycle. Chronic paronychia usually appears in adults whose occupations regularly contact water and children because of the thumb sucking habit [17].

Distal candida nail infection manifests as subungual hyperkeratosis along with onycholysis. Differentiating the clinical manifestation with dermatophytosis can be challenging, however the candida results in less extent damage to the nail compared to dermatophyte. In addition, the predilection of CO usually affects the fingernails, while most dermatophytes invade the toenails. The prevalence of distal candida nail infection is infrequent and most of the cases are related to vascular problems, such as Raynaud's phenomenon [17].

Total dystrophic onychomycosis occurs in patients with chronic mucocutaneous candidosis. The organism invasion on the nail plate results in hyperkeratotic and gross thickening of the nail. Chronic mucocutaneous candidosis has multifaceted etiology which results in the weakened cell-mediated immunity. The variety of clinical appearance depends on the severity of immunosuppression; however, thickening of the nails can be observed in advanced cases due to the Candida granuloma. In addition, the involvement of the mucous membrane is nearly presented in most cases [17].

Secondary candida onychomycosis results because of other diseases involving the nail apparatus, most commonly psoriasis [17].

## 6. Diagnostic tests

Common tests utilized in the diagnosis of onychomycosis are potassium hydroxide (KOH) preparation, fungal culture, histopathology, polymerase chain reaction (PCR), and flow cytometry (**Table 4**). The combination tests are usually performed; however, the gold standard of diagnostic tests are microscopy and culture [3].

Onychoscopy can also be used for initial diagnosis of onychomycosis. The most common findings in onychomycosis are jagged edge with spikes of the proximal part of the onycholysis, parallel bands of various color resembling aurora borealis pattern, and ruin appearance at the subungual part [3].

KOH microscopy and fungal culture are presently the gold standards to establish the diagnosis of onychomycosis. However, it remains questionable because KOH microscopy demonstrates a false-negative rate between 5% to 15% and false-positive for evaluating the medication, given that KOH microscopy visualizes both

| Test | Procedure | Pros | Cons | Fungal viability | Fungal identify |
|---|---|---|---|---|---|
| **Potassium hydroxide (KOH)** | Dissolved large keratinocytes result in the flattening of nail segment and decreasing reflection from cell borders. Examined with microscopy | Quick, on-the-spot | Low sensitivity | No | No |
| **Fungal culture** | Cleaned and clipped subungual debris of the nail are scraped into the gauze. Culture developed in the agar with or without cycloheximide. Examined with microscopy | Precise | Results obtained in ≥3 weeks, high false-negative rate | Yes | Yes |
| **Histopathology** | Stained by hematoxylin and eosin to depict the elements of the fungi. Periodic acid-Schiff or Grocott's methenamine-silver can be utilized to enhance the appearance of hyphae | Validate the presence of fungus | Involves specific laboratory equipment | No | No |
| **PCR** | Employ a target gene part of ribosomal DNA or chitin synthase genes | Quick, 48 hours | Costly | Yes (real-time PCR) | Yes |
| **Flow cytometry** | Employ granulosity, cell volume, DNA, and protein markers to produce definite profiles for fungi | Very specific | Involves great sample size, costly | No | Yes |

*Cited as is from Gupta et al. [3].*

**Table 4.**
*Diagnostic tests for onychomycosis [3].*

live and dead hyphae which are identical through microscope. Furthermore, fungal culture has a wide-ranging sensitivity from 30% to 57% and requires incubation for weeks. Latest studies comparing a variety of diagnostic tests indicate that histopathology staining has higher sensitivity than KOH microscopy or culture, although another study suggests PCR for a quicker and precise alternative for fungal culture, particularly in NDM onychomycosis. Therefore, the combination of diagnostic tests is recommended to diagnosis onychomycosis accurately. A feasible option can be a KOH microscopy and PCR (or culture in a resource-limited setting) if the KOH shows positive results [3].

In the case of CO, obtaining sample for KOH microscopy and culture can be performed from the proximal and lateral parts of the nail. Nevertheless, sample can be obtained from the distal part in the case of onycholysis. Culture result may reveal creamy-whitish colonies on Sabouraud dextrose agar media or primary isolation can also be attained using chromogenic media, for instance CHROMagar Candida®. Anti-fungal susceptibility should be performed following the identification of the isolated strains to achieve the most effective therapy. Histopathological results of

CO usually display hyphae and pseudomycelia on the nail through Schiff's periodic acid stains or Grocott's methenamine silver stains. PCR can also be utilized for further identification [2].

## 7. Treatment algorithm

Defining the resolution of onychomycosis can be achieved through clinical, mycological, and complete cure. Clinical cure is described as 100% improvement depicted by clear nail, while mycological cure is described as negative KOH microscopy and negative fungal culture, respectively. Ultimately, complete cure comprises 100% clear nail and mycological cure. The goal of treating onychomycosis for physicians and affected stakeholders are achieving the complete cure. However, it is difficult for an infected nail to return into an utterly normal appearance, particularly in advanced stage although mycological cure has been attained [3].

The treatment choices (**Figure 1**) available for managing onychomycosis are oral medication, topical therapy, and devices. Oral antifungals (**Table 5**) are the first-line therapy because they result in high success rates. Nevertheless, oral antifungals are contraindicated in patients with chronic or active liver disease, congestive heart failure, and kidney failure. Besides, oral antifungals may interact with other pharmacological agents, which can trigger a severe adverse reaction. These setbacks urged the request for the safer option which leads to the awareness of topical therapy. Topical treatments (**Table 6**) are indicated in mild–moderate cases and patients with contraindication for oral antifungals. However, they also have limitations which are smaller cure rate, prolonged therapy and difficulty applying for patients with mobility problems. Ultimately, lasers are FDA-approved device therapy for short-term clearance and/or nail enhancement. However, laser therapy is lacking conclusive guidance and its efficacy demonstrates notable disparities among all treatment modalities. Topical antifungals eradicate the fungus from the outward penetrating the dorsal part of the nail, whereas oral antifungals eliminate from the inward infiltrating the ventral part of the nail [3].

Another proposed treatment algorithm is based on the severity of onychomycosis assessed with SCIO (**Table 7**) [12].

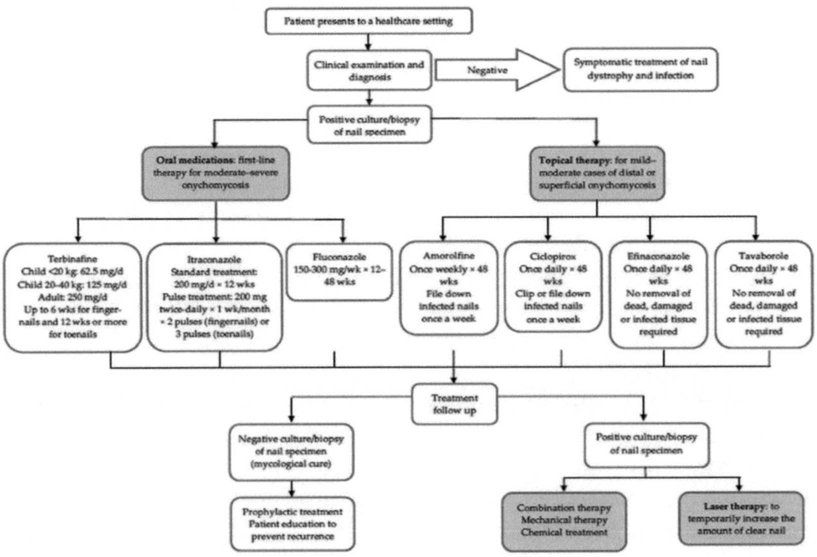


**Figure 1.**
*Treatment algorithm of onychomycosis [16]. Cited as is from Christenson et al. [16].*

| Drug name | | Terbinafine | Itraconazole | Fluconazole |
|---|---|---|---|---|
| Trade name | | Lamisil | Sporanox | Diflucan |
| Chemical structure | | Allylamine | Triazole | Triazole |
| Molecular formula | | $C_{21}H_{25}N \cdot HCl$ | $C_{35}H_{38}Cl_2N_8O_4$ | $C_{13}H_{12}F_2N_6O$ |
| Mass (g/mol) | | 291.3 | 705.64 | 306.27 |
| Mechanism of action | | Squalene epoxidase inhibitor | Lanosterol 14α-demethylase inhibitor | Lanosterol 14α-demethylase inhibitor |
| CYP[+] inhibition | | CYP2D6 | CYP3A4 | CYP2C9, CYP2C19, CYP3A4 |
| Spectrum of action | | Dermatophytes, some activity against NDMs | Dermatophytes, NDMs, and *Candida* spp. | Dermatophytes, some NDMs, and *Candida* spp. |
| Efficacy | MC | 70% | 54% | 47–62% |
| | CC | 38% | 14% | 28–36%* |
| Approval | | US–1996 | US–1995 | US–1990† |
| | | EU–1991 | EU–1989 | EU (UK)–1988 |
| | | Canada–1993 | Canada–1993 | EU (Finland)–1993 |
| | | | | China–1993 |
| | | | | Canada–1990* |
| FDA pregnancy class | | B | C | D |

*CYP, cytochrome P450; NDM, non-dermatophyte molds; MC, mycological cure; CC, complete cure.*
**Data provided are clinical cure rates.*
*†Fluconazole was FDA-approved for use in humans in 1990, but is not yet approved for treatment of onychomycosis in the US or Canada.*
*Cited as is from Gupta et al. [3].*

**Table 5.**
*Summary of available oral antifungal [3].*

| Drug name | Efinaconazole | Tavaborole | Ciclopirox | Amorolfine |
|---|---|---|---|---|
| Trade name | Jublia | Kerydin | Penlac | Loceryl |
| Chemical structure | Triazole | Oxaborole | Hydroxypyridone | Morpholine |
| Molecular formula | $C_{18}H_{22}F_2N_4O$ | $C_7H_6BFO_2$ | $C_{14}H_{24}N_2O_3$ | $C_{21}H_{35}NO$ |
| Mass (g/mol) | 348.39 | 151.93 | 207.27 | 317.51 |
| Mechanism of action | Lanosterol 14α-demethylase inhibitor | Aminoacyl †RNA synthetase inhibitor | Chelation of polyvalent heavy metal ions | $\Delta^{14}$-sterol reductase and cholestenol |
| | | | | Δ-isomerase inhibitor |
| Spectrum of action | Dermatophytes, NDMs, and *Candida* spp. | Dermatophytes, NDMs, and yeasts | Dermatophytes, *Candida* spp., and some NDMs, gram-positive and negative bacteria | Dermatophytes, NDMs, and yeasts |

| Drug name | | Efinaconazole | Tavaborole | Ciclopirox | Amorolfine |
|---|---|---|---|---|---|
| Efficacy | MC | 53.4–55.3% | 31.1–35.9% | 29–36% | 60%[125] |
| | CC | 15.2–18.8[122] | 6.5–9.1%[121] | 5.5–8.5%[123] | |
| Approval | | US–2014 | US–2014 | US–1999 | EU–1991 |
| | | Canada–2013 | | Canada–2004 | Australia–1996 |
| | | Japan–2014 | | | |
| FDA pregnancy class | | C | C | B | Poor systemic absorption, safe in animals, no studies in pregnant women† |

*CC, complete cure; MC, mycological cure; NDM, non-dermatophyte molds.*
*†Not approved by FDA, thus no pregnancy classification.*
*Cited as is from Gupta et al. [3].*

**Table 6.**
*Summary of available topical antifungal [3].*

| SCIO | Treatment approach |
|---|---|
| 1–3 | Topical treatment: remove (cut or scrape off) affected marginal parts of the nail |
| | Use topical antifungals until healthy nail regrows |
| 3–6 | Topical treatment with lower success, which often depends on growth rate |
| | Systemic therapy recommended in slower-growing nails or proximal onychomycosis type |
| 6–9 | Systemic therapy. Use scheme proposed for fingernails (e.g., itraconazole: 2 pulses of 200 mg bid) |
| 9–12 | Systemic therapy. Use scheme proposed for toenails (e.g., itraconazole: 3 pulses of 200 mg bid) |
| 12–16 | Systemic therapy. Use scheme proposed for fingernails with any antifungal (e.g., 4–5 pulses of itraconazole, 200 mg bid) |
| 16–20 | Combination therapy (systemic antifungal + topical measures) |
| | Adequate keratolytic treatment recommended |
| 20–30 | Consider nail avulsion (e.g., with urea paste), continue with systemic therapy |

*Cited as is from Sergeev et al. [12].*

**Table 7.**
*Proposed treatment approach based on scoring clinical index of onychomycosis (SCIO) [12].*

## 8. Prevention and education

As CO commonly recurs with overall onychomycosis recurrence rate of 10–53%, additional measures should be implemented to prevent this recurrence. For the clinicians, it is imperative to confirm the diagnosis and identify the infectious agent before providing treatment. Assessing and treating the comorbidities is also crucial since some comorbidities are risk factors for onychomycosis, also portend as poor prognostic factors. Tinea pedis should be treated properly as the infected skin can play a role as reservoir for the pathogens [3].

When the patients are diagnosed, the clinicians should provide them with optimal onychomycosis therapy, provide counseling on the expectations and adherence to treatment. The patients should also be provided with information to maintain hand and foot hygiene, avoid occlusive shoes, trim the nails regularly, use broad toed shoes with absorbent materials, and avoid barefoot walking in locations with abundant fungal density (e.g., swimming pool, communal showers, gymnasium floors). Good sanitization measures should be taken for previous infected socks and shoes. Socks should be washed with hot water (60 °C) for 45 minutes and shoes should be exposed to ultraviolet rays or ozone or can be sprayed with antifungal sprays. The close contacts or family members of the patients should be examined and treated if they suffer from onychomycosis or tinea pedis [3, 18].

Prophylaxis can be considered for patients with high probability to suffer from recurrence. Topical antifungal agent in the form of solution or lacquer can be applied once daily for a month then twice weekly for at least two years after the cure have been achieved [3].

## 9. Conclusions

CO is a common nail infection affecting people worldwide. Establishing the diagnosis of CO becomes a challenge for the clinicians since *Candida* spp. is a well-known normal flora inhabiting human's skin, nails and mucosa. In addition to confirming the diagnosis, the clinicians should pay attention to patient's characteristics, nail features, and the infectious agent as it can portend as poor prognostic factors. A proper treatment course along with additional measures will aid the patient to achieve complete cure and prevent recurrence.

## Conflict of interest

The authors declare no conflict of interest.

## Appendices and nomenclatures

| | |
|---|---|
| CMC | Chronic Mucocutaneous Candidiasis |
| CO | Candida Onychomycosis |
| HIV | Human Immunodeficiency Virus |
| KOH | potassium hydroxide |
| NDMs | Non-Dermatophyte Molds |
| PCR | Polymerase Chain Reaction |
| SCIO | Scoring Clinical Index of Onychomycosis |
| TDO | Total Dystrophic Onychomycosis |

## Author details

Sandra Widaty*, Eliza Miranda and Caroline Oktarina
Department of Dermatology and Venereology, Faculty of Medicine Universitas Indonesia-dr. Cipto Mangunkusumo National Central General Hospital, Jakarta, Indonesia

*Address all correspondence to: sandra.widaty@gmail.com

## References

[1] Gupta AK, Versteeg SG, Shear NH. Onychomycosis in the 21st Century: An Update on Diagnosis, Epidemiology, and Treatment. J Cutan Med Surg. 2017;21(6):525-39.

[2] Andrés T-S, Alexandro *B. candida* Onychomycosis: an Old Problem in Modern Times. Current Fungal Infection Reports. 2020;14(3):209-16.

[3] Gupta AK, Stec N, Summerbell RC, Shear NH, Piguet V, Tosti A, et al. Onychomycosis: a review. J Eur Acad Dermatol Venereol. 2020;34(9):1972-90.

[4] Widaty S, Miranda E, Bramono K, Menaldi SL, Marissa M, Oktarina C, et al. Prognostic factors influencing the treatment outcome of onychomycosis Candida. Mycoses. 2020;63(1):71-7.

[5] Gupta AK, Taborda VBA, Taborda PRO, Shemer A, Summerbell RC, Nakrieko KA. High prevalence of mixed infections in global onychomycosis. PLoS One. 2020;15(9):e0239648.

[6] Bitew A, Wolde S. Prevalence, Risk Factors, and Spectrum of Fungi in Patients with Onychomycosis in Addis Ababa, Ethiopia: A Prospective Study. J Trop Med. 2019;2019:3652634.

[7] Gregoriou S, Mpali N, Vrioni G, Hatzidimitriou E, Chryssou SE, Rigopoulos D. Epidemiology of Onychomycosis in an Academic Nail Unit in South Greece during a Three-Year Period. Skin Appendage Disord. 2020;6(2):102-7.

[8] Sigurgeirsson B. Prognostic factors for cure following treatment of onychomycosis. J Eur Acad Dermatol Venereol. 2010;24(6):679-84.

[9] Lipner SR. Prognostic Factors in Onychomycosis Treatment. Journal of Infectious Diseases and Therapy. 2014;03(01).

[10] Ranawaka RR, de Silva SH. Factors influencing cure rates of non-dermatophyte mold and Candida onychomycosis: analysis of outcomes in 81 patients who completed treatment. Int J Dermatol. 2017;56(2):202-8.

[11] de Berker D. Nail anatomy. Clin Dermatol. 2013;31(5):509-15.

[12] Sergeev AY, Gupta AK, Sergeev YV. The Scoring Clinical Index for Onychomycosis (SCIO index). Skin therapy letter. 2002;7 Suppl 1:6-7.

[13] Baran R, Hay RJ, Garduno JI. Review of antifungal therapy and the severity index for assessing onychomycosis: part I. J Dermatolog Treat. 2008;19(2):72-81.

[14] Carney C, Tosti A, Daniel R, Scher R, Rich P, DeCoster J, et al. A new classification system for grading the severity of onychomycosis: Onychomycosis Severity Index. Archives of dermatology. 2011;147(11):1277-82.

[15] Ramage G, Mowat E, Jones B, Williams C, Lopez-Ribot J. Our current understanding of fungal biofilms. Critical reviews in microbiology. 2009;35(4):340-55.

[16] Chandra J, Kuhn DM, Mukherjee PK, Hoyer LL, McCormick T, Ghannoum MA. Biofilm formation by the fungal pathogen *Candida albicans*: development, architecture, and drug resistance. Journal of bacteriology. 2001;183(18):5385-94.

[17] Ameen M, Lear JT, Madan V, Mohd Mustapa MF, Richardson M. British Association of Dermatologists' guidelines for the management of onychomycosis 2014. Br J Dermatol. 2014;171(5):937-58.

[18] Lipner SR, Scher RK. Onychomycosis: Treatment and prevention of recurrence. J Am Acad Dermatol. 2019;80(4):853-67.

Chapter 6

# *Candida albicans* and Abortion

*Humam Kasem Hussein*


## Abstract

An abortion that occurs spontaneously is known as a miscarriage. Various effectors associated with abortion such as Genetic and uterine anomalies, Endocrinopathy, immunological dysfunctions, infectious agents, environmental contaminants, psychogenetic elements, and endometriosis. Maternal infections considered the main reason for pregnancy wastage in females with Bad Obstetric History (BOH). *Candida albicans* is a dimorphic fungus that can grow as yeast or filamentous cells and considered one of the limited species of the *Candida* genus that cause humans candidiasis. It is an opportunistic fungus that responsible for mucosal infections in the mouth and genital tract. Excessive growth of *C. albicans* will follow with Vulvovaginal candidiasis (VVC). The incidence of VVC combined with chronic recurrent candidiasis is high in pregnancies than in healthy women. Several scientific researches showed the significance of VVC as an inducer of abortion, candida chorioamnionitis, subsequent preterm delivery, and immunosuppression.


**Keywords:** *Candida albicans*, Opportunistic fungi, Spontaneous abortion, VVC, Candidemia

## 1. Introduction

Mycoses considered as most ancient infections, established by Hippocrates and Galen. The fungal infection may be acute, chronic, superficial, or deep [1]. Every year, invasive candidiasis infects about 250,000 persons around the world, which leads to more than 50,000 deaths [2]. In the 19th century, mycoses fixed as infections of newborns and the genital tract in gestation and how affected by each other. The vaginal infections that resulted from yeast-like fungi of the *Candida* genus are the main infections during pregnancy [3]. Even with the presence of placenta and fetal membranes as protective sheets of him against infections, the embryo maybe infected with fungi via ascending (from the vagina) or hematogenic routes in exceptional cases. *Candida albicans* can cross that barrier without damage to the membranes. Watching the placenta plays an important role in the diagnosis of congenital candidiasis [4]. Many infants born at the 23rd week of gestation in serious conditions with congenital candidiasis and the invasion of the membranes by *C. albicans*. Intrauterine infection with *C. albicans* leads to the raising of inflammatory parameters in maternal blood (leukocytes, C-reactive protein, procalcitonin) that also detected in the child blood after delivery. So, early termination of pregnancy becomes prefers [5].



IntechOpen

## 2. Abortion

Termination of the gestation by removing fetus or embryo prior to gaining the ability to survive outside the uterus is called abortion. Nevertheless, if this process happens after the fetus acquires this ability, then it is termed a "late termination of pregnancy". If the abortion occurs spontaneously, it is termed a miscarriage. In addition, it is titled an induced abortion or "induced miscarriage" if it resulted purposely [6].

Induced abortion does not raise the risk of mental or physical complications if it ensues under legal and secure conditions [7]. Every year nearly 56 million abortion cases happen worldwide [8], half of these cases ended unsafely [9]. Unsafe abortion is considered one of the main challenges of public health in Africa and Middle East areas. In 2003, 1.5 million abortions occur in these regions in unhygienic and unexperienced conditions according to World Health Organization (WHO). From those abortions, 11% of the cases were ended with maternal death. Increasing family planning and birth control make the rate of abortion decline and that what happened in the last two decades globally [10].

In general, the causes of miscarriage are different. Several factors that can form a high degree of risk on pregnancy have been recognized. Health and medical causes have a high rate of incidence in recurrent than in spontaneous miscarriages. Cytogenetic abnormalities are probable reasons for miscarriage particularly earlier to the 9th week of gestation. Autosomal trisomies are the most common chromosomal abnormalities then 45X and triploidy. Gene inactivation in the 4 to 8 cell stages karyotype supposed to be responsible for the non-recognized cases of abortion at an earlier period of gestation.

Miscarriage also occurs by anomalies in the uterus configuration such as the bicornuate and septate uterus, which consider as congenital defects. In addition, submucosal or intramural myomata may lead to early miscarriage [11]. Occasionally, women with spontaneous miscarriage may have endocrine and autoimmune irregularities. The danger of miscarriage will increase in pregnancies who suffered from Hypothyroidism and Polycystic Ovarian Syndrome (PCO). In addition, those with low control on their blood glucose level especially in insulin-dependent diabetes mellitus [12]. The incidence of miscarriage will upsurge with the progression of maternal age. The rate of recurrence increased from 12% before 25 years to 18% after 39 years. At higher ages, anembryonic pregnancies are frequently prevalent. Menarche and menopause are the main factors that influenced maternal age. Social, economic and, cultural situations also affect the preferred family size and period between gestations [13].

Besides smoking, exposure to environmental tobacco smoke (also called passive smoking) holds the same possibility of abortion's occurrence [14]. Consumption of alcohol, caffeine also described as a weak and debatable risk factor of pregnancy loss [15]. Employments with high levels of stress are associated with spontaneous abortion [16]. Miscarriage also resulted from genital infections. *Mycoplasma hominis* and *Chlamydia trachomatis* are established as inducing factors of miscarriage existence. Pregnant women with bacterial vaginosis may be exposed to the risk of late miscarriage [17]. Bacterial vaginosis may follow deficient in lactobacilli with overgrowth of anaerobic bacteria, as well as *Mycoplasma genitalium* and *Gardnerella vaginalis* [18]. Primary infection with genital herpes will increase the risk of miscarriage existence. In addition, other infectious agents such as *Rubella*, Toxoplasmosis, *Cytomegalovirus* and, Listeriosis also fixed as probable causes of miscarriage. *Candida* species are the second most common cause of vulvovaginitis worldwide and *C. albicans* is the most common and clinically significant species that cause vulvovaginal candidiasis. Untreated vaginal candidiasis may lead to a pelvic inflammatory illness that scar the fallopian tube followed by infertility.

### 2.1 *Candida albicans*

A dimorphic fungus that can grow as yeast or filamentous cells and considered one of the limited species of the *Candida* genus that cause humans candidiasis [19]. 50–90% of all cases of humans' candidiasis are result from *C. albicans* [20]. Systemic fungal infections (fungemia) caused by *C. albicans* appeared as significant foundations of morbidity and mortality in immunocompromised patients (e.g., AIDS, cancer chemotherapy and, bone marrow transplantation). Today, hospital-acquired candidiasis became a source of major health anxieties.

*Candida albicans* is a common human flora that noticed in the gastrointestinal tract of 40% of healthy adults [21]. It is commonly a commensal creature, nonetheless, it can turn out to be pathogenic in immunocompetent individuals under various conditions. Candidiasis also can happen due to excessive growth of the fungus, which recurrently detected in immunocompromised cases including HIV-infected patients. It usually befalls the mucous membranes of the mouth or vagina in addition to a number of other parts of the body [22].

#### *2.1.1 Fungal genome*

The genome of *C. albicans* characterized by numeric rearrangements of chromosomal structures leads to creating genetic rearrangements called chromosome length polymorphisms, reciprocal translocations, and chromosome deletions. These karyotypic modifications followed by changes in the phenotype, which consider a fungal strategy of adaptation. Two species of candida (including *C. albicans* and *C. tropicalis*) have an uncommon trait in which the CUG codon, which usually specifies leucine, in these species it encodes serine. The main feature of *C. albicans* genome is extremely dynamic, and this changeability is a higher benefit for molecular, epidemiological, and population researches for this species [23].

#### *2.1.2 Heterozygosity*

The heterozygosity of the Candidal genome surpasses that persist in other genomes is common among clinical isolates. Two proteins ensued via single-base polymorphisms vary in one or more amino acids will provide the functional variances of each protein. Therefore, this condition significantly raises the number of diverse proteins encoded by the candidal genome [24].

#### *2.1.3 Biology of* Candida albicans

Candidal colonies seem large, round, white, or cream that emanates a yeasty odor on agar plates at room temperature when grown *in vitro* [25]. By fermentation process, *C. albicans* consumes; glucose and maltose and produce acid and gas, sucrose to acid, but does not ferment lactose, this was a benefit in distinguished it from other *Candida* species. Recently, molecular phylogenetic researches confirm a polyphyletic character in the genus *Candida*. Previously, most yeast that isolated from infected individuals regularly called *Candida* even in absence of a clear indication of relationship to other *Candida* species until the development of molecular methods. For example, three species of candida which are *C. guilliermondii*, *C. glabrata* and *C. lusitaniae* were misclassified and positioned in other genera until the evolution of phylogenetic reorganization [23].

### *2.1.4 Epidemiology*

Many regions of the human body like skin and mucosal surfaces are inhabited by numerous candidal species, this colonization carried a commensal nature with the host. The immune condition of the individual plays an effect on the severity of the Candida infections, so, any disturbance in immunity increase the percentage of the host's illness make the host more susceptible to infection with candidiasis. In immunocompromised patients, Candidal infections create main fungal infections [26]. Generally, oropharyngeal candidiasis is the primary illness presented in those patients, because of that malnutrition developed leading to restriction of the action of the treatment [27]. These invasive infections have many challenges against public health lead to cumulative health and economic significances because of the great mortality proportions and amplified expenditure of medical care [28].

Skin, mouth, throat, genitals, and blood are the main body regions that are usually infected with candidiasis. Generally, *Candida spp* sustains as the fourth supreme isolated pathogen from bloodstream infections (BSIs). Most cases of candidaemia are caused by *C. albicans* have been associated with a high mortality rate, while the non-albicans species are responsible for about 23% of candidemia collectively with the rare incidence of mortality. Virulence of these species depends on many elements; capability of biofilms creation, the existence of teleomorph forms, therapeutic difficulty, and resistance to conventional antifungal medicines [29]. Candidal nosocomial infections determined by organ transplantations, an increase of immunosuppression cases, and the clinical procedures that required the usage of invasive devices [30].

### *2.1.5 Host predisposing factors*

Besides the commensalism interaction between Candida species and humans and the fundamental existence of it in healthy persons, recent two decades showed an unusual overgrowth in respiratory, gastrointestinal, and urinary tracts in comparison with earlier periods. Shortly after childbirth, species colonize the mucosa of the upper respiratory passages and gastrointestinal tract. Habitually, *C. albicans* exists fluently in the internal warm crinkles and fissures of the gastrointestinal tract and vaginal tract. Candidal colonization rises nearly to 30–40% during pregnancy due to disturbance of immunity, bacterial flora, and pH level variations, while about 10% of these species are found in mucosa and skin of the genitalia in men [31].

## 2.2 *Candida albicans* and pregnancy

During pregnancy, females exposed to many physiological changes. Gestation is a complicated condition in fetal development that requires various essential substances such as glucose, fatty acids, amino acids, minerals, and vitamins. These nutrients must continuously apply to improve the process of fetal growth and to protect the health condition of pregnant women. Many pathogens that responsible for several sexual and non-sexual transmitted infections invade the women's bodies through the female genital tract (FGT), leading to vaginal infections. The common clinical symptom for female genital tract infection is vaginal discharge, which considers as the second main gynecological problem after menstrual disorders [32]. Vulvovaginal candidiasis (VVC) (also called candidal vaginitis or moniliasis) initiated by an overgrowth of candida yeast species mainly *C. albicans*. The main features of this disease are curd-like vaginal discharge, itching, erythema, burning, vulvar and vaginal irritation associated with dysuria and dyspareunia [33]. *C. albicans* overgrowth causes superficial infections such as vaginitis that are usually associated with an immuno-compromised state mucosal candidiasis. Scientific researches fixed that near to 75% of women undergo at minimum one incidence of a genital yeast infection at reproductive years of them, In addition, about 10–20% of women acquire asymptomatic vaginal colonization with Candida species during their life. While 5–10% of healthy women suffering from recurrent vaginal candidiasis without any predisposing factors. In the presentation of chronic recurrent candidiasis, pregnant women are less resistant to VVC in comparison with healthy women. The forms of infection may be acute, chronic, superficial, or deep. During pregnancy, rising in estrogen level will be followed by increasing in glycogen production in the vagina, which improves the proliferation of the yeast on the lining of it. Alterations in physiological conditions that affect the beneficial bacteria in the vagina would change the vaginal acidity reducing its pH to 5.0–6.5. This alteration in pH will increase the overgrowth of pathogenic *Candida*. Several factors such as age, menstrual cycle, sexual activity, pregnancy, and excessive use of antibiotics may lead to an increased vaginal pH [34].

Colonization of the vagina by Candida species may be enhanced by numerous factors such as pregnancy, weak immunity, obesity, diabetes, prolonged use of corticosteroids, HIV, malnutrition, consumption of high level of estrogens, Intrauterine Contraceptive Device (IUCDs), tight clothing, poor personal hygiene, intrauterine devices and diet with high carbohydrates contents. VVC is a significant infection that may lead to abortion, candida chorioamnionitis, subsequent preterm delivery, and suppression of the immune system. Even with the isolation of Other candida *spp* (*Candida tropicalis*) from aborted placenta [35], *C. albicans* considered the main one that can invade the fetal membranes. Uterus infection with candida may be occurring via the usage of IUD that might hold the yeast from contaminated external genitalia into the uterus. In many cases, the pregnancy occurs even with the presence of IUD and that may lead to candidal abortion [5]. In addition, the probability of the presence of *C. albicans* in the uterus was referred to transmit of that yeast via seminal fluid, giving some proves about the role of the male as a reservoir of *C. albicans.* This may lead to re-infection of their sexual partner besides the isolation of that yeast from the genitalia and from semen [36]. In general, the infected male stays asymptomatic carriers and that will add another difficulty to control the yeast spreading. Although its ubiquity in the vagina, intra-amniotic infection with *C. albicans* is rare and that explained the few isolates that detected from the aborted placenta [37].

## 3. Conclusion

*Candida albicans* is one of the major normal microbiota found in the human body. It converts to opportunistic microorganisms when the host underwent several physiological and pathological conditions. In pregnant women, it can reach the placenta either by cause ascending infection from the vagina or by infected seminal fluid, which may lead to abortion.

## Acknowledgements

I would like to express my sincere thanks to all researchers and scientists for their support and guidance to accomplish this work.

**Author details**

Humam Kasem Hussein
Technical Institute of Al-Najaf, Al-Furat Al-Awsat Technical University, Al-Kufa, Iraq

*Address all correspondence to: kuh.hum@atu.edu.iq

## References

[1] Parveen, N.; Munir, A.; Din, I. and Majeed, R. Frequency of vaginal candidiasis in pregnant women attending routine antenatal clinic. J. Coll. Physicians Surg. Pak. 2008; 18(3): 154-157. DOI: 03.2008/jcpsp.154157

[2] Arendrup, M. Epidemiology of invasive candidiasis. Curr. Opin. Crit. Care. 2010; 16(5): 445-452. DOI: 10.1097/mcc.0b013e32833e84d2

[3] Mendling, W.; Friese, K.; Mylonas, I.; Weissenbacher, E.; Brasch, J.; Schaller, M.; Mayser, P.; Effendy, I.; Ginter-Hanselmayer, G.; Hof, H.; Cornely, O. and Ruhnke, M. Vulvovaginal Candidosis (excluding chronic mucocutaneous candidosis). Guideline of the German Society of Gynecology and Obstetrics. 2013; 75(4): 342-354. DOI: 10.1055/s-0035-1545741

[4] Benirschke, K. and Kaufmann, P. Pathology of the human placenta. Springer-Verlag, New York, 2000. DOI: 10.1007/978-3-642-23941-0

[5] Ito, M.; Nakashima, A.; Hidaka, T.; Okabe, M.; Bac, N. and Ina, S. A role for IL-17 in induction of an inflammation at the fetomaternal interface in preterm labour. J. Reprod. Immunol, 2010; 84: 75-85. DOI: 10.1016/j.jri.2009.09.005

[6] Grimes, D. and Stuart, G. Abortion jabberwocky: the need for better terminology. Contraception. 2010; 81 (2): 93-96. DOI: 10.1016/j.contraception.2009.09.005

[7] Lohr, P.; Fjerstad, M.; Desilva, U. and Lyus, R. Abortion. BMJ. 2014; 348: f7553. DOI: https://doi.org/10.1136/bmj.f7553

[8] Sedgh, G.; Bearak, J.; Singh, S.; Bankole, A.; Popinchalk, A.; Ganatra, B.; Rossier, C.; Gerdts, C.; Tunçalp, Ö.; Johnson, B.; Johnston, H. and Alkema, L. Abortion incidence between 1990 and 2014: global, regional, and subregional levels and trends. The Lancet. 2016; 388 (10041):258-267. DOI: 10.1016/S0140-6736(17)31794-4

[9] Sedgh, G.; Singh, S.; Shah, I.; Åhman, E.; Henshaw, S. and Bankole, A. Induced abortion: Incidence and trends worldwide from 1995 to 2008. 2012; 379 (9816): 625-632. DOI: 10.1016/S0140-6736(11)61786-8

[10] Sedgh, G.; Henshaw, S.; Singh, S.; Bankole, A. and Drescher, J. Legal abortion worldwide: incidence and recent trends. Int. Fam. Plan. Perspect. 2007; 33 (3): 106-116. DOI: 10.1363/3310607

[11] Cramer, D. and Wise, L. The epidemiology of recurrent pregnancy loss. Semin Reprod. Med. 2000; 18(4): 331-339. DOI: 10.1055/s-2000-13722

[12] Regan, L. and Rai, R. Epidemiology and the medical causes of miscarriage. Baillieres Best Pract. Res. Clin. Obstet. Gynaecol. 2000; 14(5): 839-854. DOI: 10.1053/beog.2000.0123

[13] Nybo Andersen, A.; Wohlfahrt, J.; Christens, P.; Olsen, J. and Melbye, M. Maternal age and fetal loss: population based register linkage study. 2000; 320(7251): 1708-1712. DOI: 10.1136/bmj.320.7251.1708

[14] Mishra, G.; Dobson, A. and Schofield, M. Cigarette smoking, menstrual symptoms and miscarriage among young women. Aust. N. Z. J. Public Health. 2000; 24(4): 413-420. DOI: 10.1111/j.1467-842x.2000.tb01604.x

[15] Leviton, A. and Cowan, L. A review of the literature relating caffeine consumption by women to their risk of reproductive hazards. Food Chem. Toxicol. 2002; 40(9): 1271-1310. DOI: 10.1016/s0278-6915(02)00092-3

[16] Mulder, E.; Robles de Medina, P.; Huizink, A.; Van den Bergh, B.; Buitelaar, J. and Visser, G. Prenatal maternal stress: effects on pregnancy and the (unborn) child. Early Hum Dev. 2002; 70(1-2): 3-14. DOI: 10.1016/s0378-3782(02)00075-0

[17] Oakeshott, P.; Hay, P.; Hay, S.; Steinke, F.; Rink, E. and Kerry, S. Association between bacterial vaginosis or chlamydial infection and miscarriage before 16 weeks' gestation: prospective community based cohort study. BMJ. 2002; 325(7376):1334. DOI: 10.1136/bmj.325.7376.1334

[18] Larsson, P.; Bergstrom, M.; Forsum, U. Jacobsson, B.; Strand, A. and Wolner-Hanssen, P. Bacterial vaginosis. Transmission, role in genital tract infection and pregnancy outcome: an enigma. APMIS. 2005; 113(4): 233-245. DOI: 10.1111/j.1600-0463.2005.apm_01.x

[19] Erdogan, A. and Rao, S. Small intestinal fungal overgrowth. Curr. Gastroenterol Rep. 2015; 17 (4): 16. DOI:10.1007/s11894-015-0436-2

[20] Martins, N.; Ferreira, I.; Barros, L.; Silva, S. and Henriques, M. Candidiasis: predisposing factors, prevention, diagnosis and alternative treatment. Mycopathologia. 2014; 177 (5-6): 223-240. DOI :10.1007/s11046-014-9749-1

[21] Mukherjee, P.; Sendid, B.; Hoarau, G.; Colombel, J.; Poulain, D. and Ghannoum, M. Mycobiota in gastrointestinal diseases. Nat. Rev. Gastroenterol Hepatol. 2015; 12 (2): 77-87. DOI: 10.1038/nrgastro.2014.188

[22] Yehuda, Z.; Saar, B.; Estella, D.; Vadim, S.; Clariel, I. and Tamar, H. Colonization of Candida: prevalence among tongue-pierced and non-pierced immunocompetent adults. Oral Dis. 2010; 16(2): 172-175. DOI: 10.1111/j.1601-0825.2009.01618.x.

[23] Butler, G.; Rasmussen, M. and Lin, M. Evolution of pathogenicity and sexual reproduction in eight Candida genomes. Nature. 2009; 459 (7247): 657-662. DOI: 10.1038/nature08064.

[24] Larriba, G. and Calderone, R. (2008). Heterozygosity and Loss of Heterozygosity in *Candida albicans*. Pathogenic Fungi: Insights in Molecular Biology. Caister Academic Press. DOI: https://doi.org/10.21775/9781910190678

[25] Arnaud, M.; Costanzo, M.; Inglis, D.; Skrzypek, M.; Binkley, J.; Shah, P.; Binkley, G.; Miyasato, S. and Sherlock, G. (2011). CGD Help: Non-standard Genetic Codes. Candida Genome Database. http://www.candidagenome.org/help/code_table s.shtml

[26] Fidel, *P. candida*-host interactions in HIV disease: relationships in oropharyngeal candidiasis. Adv. Dent. Res. 2006; 19: 80-84. DOI: 10.1177/154407370601900116

[27] Horn, D.; Neofytos, D.; Anaissie, E.; Fishman, J.; Steinbach, W.; Olyaei, A.; Marr, K.; Pfaller, M.; Chang, C. and Webster, K. Epidemiology and outcomes of candidemia in 2019 patients: data from the prospective antifungal therapy alliance registry. Clin. Infect. Dis. 2009; 48(12): 1695-1703. DOI: 10.1086/599039

[28] Lai, C.; Wang, C.; Liu, W.; Huang, Y. and Hsueh, P. Time to positivity of blood cultures of different Candida species causing fungaemia. J. Med. Microbiol. 2012; 61(5): 701-704. DOI: 10.1099/jmm.0.038166-0

[29] Brunke, S. and Hube B. Two unlike cousins: *Candida albicans* and *C. glabrata* infection strategies. Cell Microbiol. 2013; 15(5):701-708. DOI: 10.1111/cmi.12091

[30] Wächtler, B.; Citiulo, F.; Jablonowski, N.; Förster, S.; Dalle, F.; Schaller, *M. candida* albicans–epithelial

interactions: dissecting the roles of active penetration, induced endocytosis and host factors on the infection process. PLoS. One. 2012; 7(5):1-10. DOI: 10.1371/journal.pone.0036952

[31] Kim, J. and Sudbery, *P. candida* albicans, a major human fungal pathogen. J. Microbiol. 2011; 49(2):171-177. DOI: 10.1007/s12275-011-1064-7

[32] Akinbami, N.; Babalola, O.; Shittu, O.; Tijani, M. and Adekola, A. Detection and Epidemiology of Vulvovaginal Candidiasis among Asymptomatic Pregnant Women Attending a Tertiary Hospital in Ogbomoso, Nigeria. Int. J. Biomed. Res. 2015; 6 (7): 18-23. DOI: https://doi.org/10.7439/ijbr.v6i7.2242

[33] Rathod, S.; Klausner, J.; Krupp, K.; Reingold, A. and Madhivanan, P. Epidemiologic Features of Vulvovaginal Candidiasis among Reproductive- Age Women in India, Hindawi Publishing Corporation. Infect. Dis. Obstet. Gynecol. 2012; 1(8): 42-45. DOI: https://doi.org/10.1155/2012/859071

[34] Gonzalez, M.; Elizondo, M. and Ayala, J. Trends in Species Distribution and Susceptibility of Blood stream Isolates of Candida collected in Monterrey Mexico to Seven Antifungal Agents, J. Clin. Microbiology. 2008; 46(9): 2902-2905. DOI: 10.1128/JCM.00937-08

[35] Hussein, H. K. Isolation and Detection of *Candida tropicalis* from Aborted Placenta in Al-Najaf city/Iraq. International Journal of Pharmaceutical Quality Assurance. 2018; 9(2); 204-207. DOI: 10.25258/ijpqa.v9i2.13648.

[36] Horowitz, B.; Edelstein, S. and Lippman, L. (1987). Sexual transmission of Candida. Obstet. Gynecol. 69(6):883-886.

[37] DiGiulio, D. Diversity of microbes in amniotic fluid. Semin Fetal Neonatal Med. 2012; 17(1): 2-11. DOI: 10.1016/j.siny.2011.10.001